In Love With Betty The Crow

In Love With Betty The Crow

THE FIRST 40 YEARS OF ABC RN's
THE SCIENCE SHOW

ROBYN WILLIAMS

ABC
Books

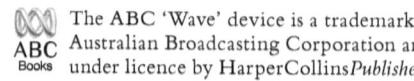 The ABC 'Wave' device is a trademark of the Australian Broadcasting Corporation and is used under licence by HarperCollins*Publishers* Australia.

First published in Australia in 2016
by HarperCollins*Publishers* Australia Pty Limited
ABN 36 009 913 517
harpercollins.com.au

Copyright © Robyn Williams 2016

The right of Robyn Williams to be identified as the author of this work has been asserted by him in accordance with the *Copyright Amendment (Moral Rights) Act 2000*.

This work is copyright. Apart from any use as permitted under the *Copyright Act 1968*, no part may be reproduced, copied, scanned, stored in a retrieval system, recorded, or transmitted, in any form or by any means, without the prior written permission of the publisher.

HarperCollins*Publishers*
Level 13, 201 Elizabeth Street, Sydney, NSW 2000, Australia
Unit D1, 63 Apollo Drive, Rosedale, Auckland 0632, New Zealand
A 53, Sector 57, Noida, UP, India
1 London Bridge Street, London, SE1 9GF, United Kingdom
2 Bloor Street East, 20th floor, Toronto, Ontario M4W 1A8, Canada
195 Broadway, New York, NY 10007

National Library of Australia Cataloguing-in-Publication data:

Williams, Robyn, 1944- author.
 In love with Betty the crow : the first 40 years of ABC RN's the science show / Robyn Williams.
 ISBN: 978 0 7333 3501 3 (paperback)
 ISBN: 978 1 4607 0623 7 (epub)
 Subjects: Science show (Radio program)
 Scientists – Australia – Interviews.
 Technology – Australia.
 Science – Australia.
791.4472

Cover design by Darren Holt, HarperCollins Design Studio
Cover image by shutterstock.com
Typeset in Trump Mediaeval by Kirby Jones

Contents

Acknowledgements	vii
Prologue	1
1 *The Science Show* No.1	17
2 Enter Norman and Crunch Time	20
3 Here Come the Women	37
4 Here Come the Animals	55
5 There Goes the Clever Country	68
6 Climate Science and Terrorism	82
7 Some Experiments in Broadcasting	96
8 Stars	106
9 Failures	138
10 Successes	147
11 In the Beginning	157
12 The Future	170
13 Is Australia Too Big?	189
14 Terra Nullius	197
15 Marine Science	208
16 Germs!	219
17 The War on Science	228
18 Here Comes the A-team	237
19 Being Co-opted Into the Elite?	259
Index	267

Acknowledgements

This is a book about a program, not about me. That is why there are lots of extracts from *The Science Show* indicating conclusions that may seem surprising, but over time may turn into received wisdom.

First accolade goes to Sharon Carleton and John Spence (ABC archives) who found all the extracts. Sharon is also a long-standing *Science Show* contributor of immense range and flair. She has never refused a request, however irksome.

David Fisher has been with me for years (he is the 'team' people sometimes refer to) and does so much that the new technology demands but which many don't notice in terms of effort: the pods, the websites, the photos, the transcripts, the program production and he has other programs (*The Naked Scientists*) to look after. David has found so many references I needed.

Others who have produced the program over the years have been Halina Szewczyk, Nicole Gaunt, Polly Rickard, Barbara Hucker, Mary Mackel and Leigh Dayton. Of the contributors Pauline Newman is a star, having been in the BBC Science Unit she is now a professor of science communication at Arizona State University and still does reports for us. Others helping over the decades have been Lynne Malcolm, Norman Swan, Joel Werner, Richard Aedy, Kirsten Garrett, Jonathan Nally, John Challis and Johnnie Merson. Peter Pockley, who founded the ABC Science Unit in 1964, provided many a brave report.

So many brilliant broadcasters have made whole *Science Shows* over the years including David Ellyard, Anne Deveson, Matthew Crawford, Anna-Maria Talas, Julie Rigg, Martin Redfern, and those who've made whole series, are acknowledged within this book.

Our technical people are different each week — gone are the days when someone was dedicated to the show. But I salute them all.

I am grateful to those at ABC Books for help in dealing with a manuscript begun the day after I left hospital following two rather shattering cancer operations. I did not, exactly, have 'chemo brain' but it felt like it. I am ordinarily a 'master of allusion', as a former editor of mine once put it. I know what I mean and too quickly assume everyone else does as well. Writing ever such succinct radio scripts on my typewriter every day doesn't help such tendencies towards terse, cryptic communication and I am delighted Lachlan and Helen felt free to tell me so.

The ABC is like a difficult lover: often infuriating but impossible to put aside. Its broadcasters are simply wonderful and I owe them every skill I possess. The scientists themselves are a never-ending inspiration and Australia is blessed, more than it deserves, with such superb ones, old and young. But our listeners are what it is all about, now all over the globe. I know it is naff to point to ratings, however, our podcast figures alone towards the end of 2015, when you combined *The Science Show* and Lynne Malcolm's *All In the Mind* (also from the Science Unit) came to over SEVEN MILLION. We must be doing something right.

Finally, I have a tendency, usually thwarted, to fall prey to fatal diseases. Norman Swan has saved me more than once, over the years, and my lady, Dr Jonica Newby, rescued me at the start of 2015 with CPR — cardiopulmonary resuscitation, powerfully delivered. Both have been magnificent through these rugged times, as have my former wife and very good friend Pamela Traynor and my children Jessica and Tom. This book is for them.

Prologue

This book is based on about 2040 programs involving 14,280 stories and 7140 professors. It has involved 110,160 minutes or 1836 hours (76.5 days) of continuous broadcasting. These figures are, of course, imprecise, but who's to tell? What it means is that the task is nigh impossible, like telling the story of a long war with 95 per cent of the process apparently dreary routine — me sitting with recordings, razors and sticky tape (or nowadays tracking green waves on screen), editing for hours — or else, the remaining 5 per cent, involving thrilling encounters with celebs discussing the meaning of life. There is no coherence about the story of a radio program, especially when spread over four decades. So this book is more a personal conversation than a solemn history.

Why am I in love with Betty the Crow?

Well, I am immensely fond of animals but this does not mean I failed as a Real Man to grow up and turn properly to apparatus, flexes and accelerators (My engineer father said, 'there is physics and all the rest is stamp collecting'). Biology is now every bit as sophisticated and mathematical as physics. No, animal behaviour is one main field where developments have been spectacular in recent decades, with significant ramifications, and the sight of Betty the New Caledonian crow becoming an engineer and solving problems with insight, creativity and planning makes me cry with delight. (Other birds, and plenty of dogs, have also broken boundaries.) So, you can pin them up with all those vaccines, deep dives, space shots and techno-wizardry at the cutting edge of science, feeding the addiction that has kept my colleagues and me at the scientific mike for so long.

It has all been with the ABC, but not exclusively so. The public may perceive barriers between employees of Aunty and our commercial friends, but this is illusory. At our level, we're all mates. And, over the years, how it has all changed! I started in a kind of collegial diaspora, with the ABC spread all over Sydney and the masters of the separate 'colleges' were like powerful barons with retinues, rows of tapping secretaries and very long daily commitments called 'lunch'.

Today we occupy an endless open-plan IT factory, with the barons of drama, religion, science and sport long gone while we toil away silently with unlimited schedules, looking sometimes like Charlie Chaplin in *Modern Times*, buffeted by anxiety and machinery, insatiable screens and airtimes.

Doing this, we are also giving the public a false sense of security about legacy because we are hiding what it really costs to do our jobs. We are always packing microphones to take on holidays in case an opportunity abroad materialises or an exotic location offers broadcasting possibilities. In 2014 I went on a private trip to South America and came back with two entire *Science Show*s about Galapagos and Machu Picchu. A taxpayer-funded perk? No way. When being treated for cancer a couple of weeks before I started writing this book, I interviewed Sir Philip Campbell, Editor-in-Chief of the journal *Nature*, while reclining in my hospital room. The nurses were taken aback when I made them leave. But you can't miss a story! Our successors will wonder how the hell we managed it.

* * *

Modern science is very young despite its intellectual foundations in ancient Greece. The transition from the classical approach to a profession is barely two hundred years old. This means that the role of scientists is still uncertain, evolving, hard to pin down. For this reason, I have invariably included history and biography in my reporting. Not only does it make science more human, it also gives it context.

In early 2014, as I said, I travelled to the Galapagos Islands with my family and read a lot about Charles Darwin on the way. I was struck by his age when the good ship *Beagle* first set sail; he was barely twenty-three. On the way, in Peru, I pointed out a waiter of the same age to some of my fellow travellers and they were suitably staggered by his youthfulness. But while boy Darwin was away on the *Beagle*, in 1833 at Cambridge, Darwin's alma mater, the Reverend William Whewell actually coined a name for those who did research in science. They were no longer considered to be 'natural philosophers', because they got their hands dirty and handled apparatus — even dead animals — almost 'trade', one might say! Whewell chose 'scientist', though some thought it sounded too much like 'atheist'. (By the way, the question about what to call these new professionals was put by poet Samuel Taylor Coleridge.)

The term 'journalist' was coined only three years before, in 1830, according to Professor Nicholas Lehmann, Dean Emeritus at the Columbia School of Journalism. I mention all this because I am fond of the idea that scientists and journalists were born so closely together, and I like to feel there has been some synergy between them: *truth*. As A.C. Grayling, the philosopher, put it in another context, '*Religion and science have a common ancestor — ignorance*'.

* * *

The Science Show has been afflicted with funding cuts since the beginning, but it was the very first cut, strange as it may seem, that made the program last so long with me attached.

At the time of *Science Show* No.1, in August 1975, life was spectacularly different from how it is today. There were no mobile phones, no laptops, no internet, no DVDs, ATMs or any other electro-acronyms. Television in Australia had just gained colour and jumbo jets had been around for a little while, but you could not believe, when standing beneath them, that they could ever take off.

A simpler world? Maybe. But the new program had no conscious determination to look to the future on the assumption that some revolution was coming. Any enthusiastic fulminating about technology to come sounded like mere sales hype then. There was no concept (beyond fiction or among the enlightened few) that nearly everything would change. For most, the world was too riven by Vietnam, New Age suspicions about most authority — even the excitement of moon landings (so recent) were often recast into cynical reflections on the military.

So we thought we would talk to scientists and business people, see what research they were doing and then report it — work in progress rather than anything too futuristic. I still believe that the essential ingredient in science journalism is shoe leather. You get out of the office to where the work is going on and do the expected interview. Then, if you wait a few minutes, the inevitable question comes: 'By the way, have you heard about this…?' The tip you get by being there is gold!

When the first show was transmitted from Canada, where I happened to be at a gigantic conference, all I assumed was that I would continue as before with no major plans until some juicy alternative turned up, perhaps on ABC TV, maybe outside the ABC. I'd been there only briefly.

It was the latter that transpired, although still with Aunty. I was asked to apply for an ABC Radio job in London, covering all subjects and parts of continental Europe, if the occasion arose. This was the job from heaven: an ABC flat in the centre of London next to the BBC, Paul Lyneham, John Highfield, Paul Murphy (and one day Tony Jones and Fran Kelly) to play with, anyone of calibre in the arts or sciences to chat to — for a

31-year-old with barely three years' experience in the media as a reporter it was too tempting.

I applied. And got the job. This was after barely three months of doing *The Science Show*. Then came 11 November 1975. Whitlam was sacked, Fraser won the election and, as ever, the first action of the Coalition was to cut the funds of the ABC. My London job was the first to disappear.

What to do? Well, I just kept going with this new beast, *The Science Show*. I took lines that were then current: anti-psychiatry, critiques of 'scientism' (when researchers claim to understand the whole of human progress through interpreting ant colonies or songbirds' parental schemes). We heard from the radicals of science, such as Steven Rose in Britain, then an anti-war activist exposing the hideous nature of Vietnam weaponry; Paul Ehrlich from Stanford on overpopulation (his book *The Population Bomb* was only seven years old then); but also from rather less qualified folk with dreadlocked hair and long green cardigans who waxed on about 'energies' and 'cures'. Being critical of science too easily lapsed into flirting with the fringe, as you'll see in Chapter One. May I be forgiven.

That was a long time ago. We are still critical of science, when need be — we are not cheerleaders for R&D. But the yardstick is *evidence* not attitude. And the evidence is there in abundance if you seek it out using that essential shoe leather.

And now, in 2015, I am more enraptured by science than ever. There are several reasons for this, embedded over the years.

The first is that science seeks the truth. It also has mechanisms to expose untruth. They may take a while to work, but they are far swifter this century than last. Compared with politics, law, economics and many other professions and jobs, science is a shining light of probity, even in medical research, where uncertainties abound.

Second, scientists are such pleasant, keen people. I noticed this when a panel of them were invited on *Q&A*, the ABC TV show hosted by Tony Jones. They included two of our Nobel

Prize winners, Brian Schmidt and Peter Doherty. There were no spiteful interchanges, no devious rhetoric, only enthusiasm and plain speaking. All was pleasant. These guys deal with reality — and it showed.

Third, the results scientists get can be so exciting: *Curiosity*, the rover, landing on Mars; The Hobbit, our hominid relative discovered in Flores, Indonesia; our close genetic links to Neanderthals — more surprising as time passes; the brilliance of birds; the cleverness of dogs; the revolutionary power of our technological innovations.

Fourth, they create our wealth. Call it 30 per cent or 60 per cent of GDP in countries such as ours; it is a huge contribution. Not only does science provide with its inventions, it also makes the nation more efficient by exposing inefficiencies and offering better ways — needless back-pain monitoring, unnecessary wisdom-tooth extraction, food wastes. Billions can and have been saved by means of scientific investigation of what works best.

Fifth, as we found at the very start of *The Science Show*, environment is vitally important. Science is showing how.

Can there be any other human activity with such a record? That is why, for me, forty years has been such an adventure.

1 *The Science Show* No.1
The Age of Flake

Two massive changes occurred in 1972, the year I wandered off the street into the ABC Science Unit and was hired, albeit on a temporary basis.

The first was the United Nations Conference on the Human Environment held in Stockholm. We had vague ideas of the importance of green issues but suddenly became aware that the concern was, or should be, on a global scale. This was the first of the UN talkfests we are now used to, often seeing thousands of frightfully well-appointed delegates flying in from afar and achieving very little. The great explorer from South Australia, Tim Jarvis, who attended the conference in Lima on climate in 2014, said that event was 'disgraceful'. After twenty years of trying, 10,000 people could still not produce an effective agreement. Paris 2015 has been an improvement, just.

But in 1972, the simple presence of a world conference did make a positive impact. We were more aware than ever before that the soil, trees, water and creatures were precious and utterly vulnerable. The scale of our environmental challenge was being redefined as never before.

The second change was the cancellation of the Apollo missions, meant to go on past 1972 but abandoned by President Nixon because of cost. That meant that the twenty-year excitement about science in general, not only in space, from Sputnik to Apollo 17, was over. It had been an era of experimentation, daring, the recruitment of young minds to science, and a time when we expected, nearly every day, to read in the newspapers of some other new wonderful achievement out of the blue.

With these changes came a new darker mood. The seventies proceeded with a fresh feeling of mistrust and even cynicism. 'Alternative' was OK but conventional science was sniffed at by my *gen-gen-gen*eration. Apollo was all about the US military, we told each other. By contrast, and to my surprise (almost against my will), I was thrilled by every minute of Apollos 16 and 17, which I covered live in the radio studio in 1972 as my first real radio job. 'Roger. Copy, Houston,' said the calm astronaut as he was hurtled towards the moon. Wow!

But, yes, if you looked carefully you would see that more than half the world's scientists did indeed work in military R&D, even if some of it was seemingly prosaic, like the development of new fuels and materials. Frank Barnaby, then director of the Stockholm International Peace Research Institute, would repeatedly offer evidence that more than half of our research funds were still dedicated to war. As he said on *The Science Show*:

> The amount of resources, both manpower and financial, put into military R&D is huge. The number of physicists and engineering scientists employed in military research is more than one half of the number employed in peaceful research. And the money given to military research is about 35,000 million dollars, which is much more than is given to peaceful research. So, we put much more effort into military research than we do into peaceful research, which is really a tragic state of affairs.
>
> If there were no scientists in military research, military research would not be possible. It is in this sense that the arms race is most dangerous. So I can say that scientists, by

their activities in military research, are responsible for the arms race.

What's more, these and other state-employed scientists were not allowed to talk to us. We met blockages all the time. The free ranging calling of the friendly, willing scientist was a long way off, except at some campuses and PR agencies. Indeed, at that time, if we wanted to broadcast, say, a piece on cancer, we had to write off to the Department of Health in Canberra for permission. A letter from them saying yes or no would sometimes take weeks to arrive.

We were also a little sceptical of the scientists' intellectual range. They did their narrow work, published, and then allowed the press and random know-alls to pronounce on the results. So, we were told that women were wired to be Betty Homebodies, were intrinsically labile emotionally and would benefit from unending diazepam daily doses. Or we were given ready-made tapes from embassies (Radio Moscow, The British Foreign Office) demonstrating that their new technologies would make our lives sublime. We took on these unabashed promotions with relish, putting them to air as electronic press releases — virtual ads!

Meanwhile, the quest against 'scientism' proceeded. One of our dependable critics of psychiatry was himself of the profession: R.D. Laing. He also condemned birth practices as carried out in our hospitals, saying that women, with all the shaving and medical priorities, were being forced to ignore what was a natural event. We continued with Ronnie Laing until he began to insist on a hundred pounds in cash before interviews so he could purchase whisky.

But let's go back to that first ever *Science Show*, as broadcast on 30 August 1975, and how that epitomised some of the preoccupations I've mentioned. I was in Vancouver at the huge Pacific Science Congress with about 8000 delegates and plenty of stars. I selected those at the top of the order and with powerful things to say. What is staggering is how much that first *Science Show* resonates with what matters today.

First came nuclear weapons. The world was deeply worried about the sheer number of missiles and the cliché was we could wipe out civilisation a hundred times over. Was the volume of nukes far too risky; were the odds, and potential accidents, setting us up for Armageddon?

Those questions may seem quaint now, as other concerns about the economy, refugees and climate have taken over. But it took that splendid magazine *The Economist* to set the record straight in 2015 with a cover story — 'The new nuclear age' — suggesting we are barely better off forty years later. The risks remain.

This is what William Epstein of the United Nations said back then in the first-ever interview in *The Science Show*:

> For the first time in a quarter of a century, I am getting really scared, because as the nuclear powers keep on proliferating nuclear weapons, they're going to make them smaller and [make] more and more of them.
>
> And as more and more countries acquire nuclear capability, it is almost inevitable that terrorists, politically motivated groups, are going to eventually get hold of these things and use them for blackmail, ransom. And you know, this could mean the end of civilised society as we know it, because how do you dare even turn down a blackmail request here? Whole cities and countries can be held up to blackmail by terrorists.
>
> The world has become complacent. People have been lulled and gulled into complacency. They're not making any protests. They think the thing is too difficult for them, it's beyond they're comprehension, which is a lot of rubbish. What the people have got to do is to raise hell with their own government.
>
> It's no longer a race in numbers; it's a race in technology. It's a scientific race. And as technology keeps becoming more

and more science-fiction like, and more and more bizarre, the race is going to get worse not better. This is a race to insanity and oblivion.

This was incredibly prescient: forty years ago and a glimpse of our present predicaments. But it was my interview with Lord Ritchie Calder, the energy expert from the House of Lords in Britain, which is stunning, though none of us realised this at the time. Here was a clear warning in stark language, about climate change and fossil fuels; not only forty years ago but referring to concerns expressed twenty-five years before that:

> In the course of the last century, we've put 360,000 million tons of fossil carbon into the atmosphere. On the present trends, the accumulated requirements between now and 2000 AD will come out at something like 11,000 million tons of coal a year, 200,000 million tons of crude petroleum and liquid natural gas and 50 million, million cubic metres of natural gas. Now remember, this is coming out of the bowels of the Earth. We're taking it out and we're throwing it back into the atmosphere, and into the climatic machine, the weather machine where it is beginning to affect the climate itself. Now this is a very serious matter. And to me there's no question the world's climate has changed.
>
> You will simply be confronted by a situation which will make life virtually intolerable.
>
> We were very emphatic in 1963. We were talking at the Rome conference, the UN conference, on new sources of energy ... There is nothing that we're now discussing with such alarm and despondency that we weren't discussing over the last twenty-five, thirty years.

So here was a warning, in August 1975, from one of the world's foremost energy experts, a person who spoke on the issue in the House of Lords, that climate was changing and we should act accordingly with a sense of urgency. One of the paradoxes of that piece is that years later Ritchie Calder's son, the renowned BBC Science documentary producer Nigel Calder, would turn up in the film *The Great Global Warming Swindle* deploring climate science and opting instead for a theory espoused by the excitable Dane Dr Henrik Svensmark, that the warming, as part of a natural cycle, is caused by cosmic rays and cloud formation. The evidence for this view is thin.

Then, in that *Science Show* No.1, came forests, wildlife and the main realisation of the new green concerns. Our speaker was Dr Ian McTaggart-Cowan, a scholar of the highest accomplishments and president of the conference I was attending, the Pacific Science Congress. Dr McTaggart-Cowan's concerns give a hint of what was to come from Dr David Suzuki, then a professor of genetics at that very same campus of UBC, the University of British Columbia.

> An examination of the rates at which our larger vertebrates are being reduced in numbers to the extinction point reveals an alarming acceleration. Of a total of 594 species, 44 per cent of them have left this world as a result of direct overkill by man; 57 per cent of them have become extinct as a result of our massive alterations of the environment; 27 per cent have perished as a result of our introduction into their habitats of new animal hazards. These are the major sources of extinctions.
>
> This analysis reveals that the dominant forces acting to endanger the survival of reptiles and mammals have been overkill by man. The greatest concentrations of endangered species to be found within single political jurisdictions today are in Hawaii, in New Zealand and Galapagos. In no one of these areas is it yet certain that further declines and even further exterminations can be avoided.
>
> I see few signs of substantial change in present attitudes towards wildlife and wild lands that could permit an

encouraging forecast of the end to the steady erosion of the world's biota that is leading to a world of ever less variety, a less suitable environment for man. World movements towards conservation of the living environment are better organised now than ever before. In many governments there are agencies charged with maintaining a healthy and vital environment. Nevertheless, the continuing increase in populations of many areas show signs of diverting our attention from the vitally important long-term priorities of conserving wild living resources and focussing them on short-term issues.

When serious priority decisions arise, the governments of these nations still regard development as a top priority and, intrinsically by definition, good. The unaltered biosystem is still seen largely as a challenge to development.

The piecemeal destruction of wetlands I see going on everywhere. I see estuaries used as garbage dumps in the most enlightened countries in the world; the conversion of native forest into vast areas of exotic monoculture; the search for new energy sources using unnecessarily destructive techniques; the gradual erosion of marginal land by inappropriate agriculture; the destruction of fragile lands by so-called recreational use of motorcycles and other all-terrain vehicles. These are the insults on our biosphere of affluent societies.

A warning, forty years ago, of what was to come. Not by a fiery greenie but by a steady, responsible leader of research. Dr McTaggart-Cowan went on to become vice chancellor at the University of Victoria in Canada and died just shy of his hundredth birthday. His words resonate terrifyingly today. Ninety per cent of fish are now gone from the sea; there are estimates that two-thirds of all animal species may be extinct by 2100; one in eight birds will have disappeared quite soon.

In the rest of this first program we heard from futurologist Herman Kahn of the Hudson Institute saying growth was necessary and the environmental cautions should be borne in mind. He also warned that Australia should expect its earnings from commodities to plunge and that we should therefore invest more widely. Then there was superstar Thor Heyerdahl still banging on about canoes sailing from Peru to populate the Eastern Pacific — a romantic thought now obliterated by human genome tracking. It was interesting how resistant Heyerdahl was to debate. I broadcast a fulminating Australian anthropologist (Professor Peter White from the University of Sydney recorded in a university lavatory because the acoustics were good) condemning Heyerdahl for refusing to answer questions after his lecture and trying to perpetuate a view of Polynesian migration then and now thoroughly disputed.

How much in that first program could be put to air today? I suspect no one would find any of it exceptional. It is, perhaps, shocking that we have not done enough to ameliorate the crises in weapons, climate, extinctions and population. There is some progress, but does it match the challenge?

* * *

There were no obvious signs of flake (beyond Peru) in that first show. Most journals and major science conferences have always tried to keep wingnuts away from their deliberations. But I found plenty on the way to Vancouver via California as I collected material for the very first season of programs. The extracts below still cause me some embarrassment, but I learned my lesson later and made amends. The point about fringe material is that it is so easy. Much of the public is only too willing to believe in magic, and the perpetrators are keen to use whatever purple language it takes, as in the case of Professor William Tiller.

I cannot remember how I came to meet Tiller. Perhaps he was famous and the Stanford University people simply put him on a list. Stanford is one of the very top outfits, I say in my defence, but even so ... During my interview I was slightly bothered by the unrestrained gush he offered, but more excited by his purple unabashed overstatement. Here were headlines — and they were cheap! He was especially keen on psychokinesis, moving objects by just thinking about them, healing by the laying on of hands and ESP. He called his revelations 'The Greatest Revolution Since The Greeks'!

> One of the most dramatic [examples] is in the area of psychokinesis, the ability to move objects with the mind. I was invited to go to the Soviet Union in 1971, as a member of a seven-man investigative team to observe what people were doing in the psycho-energetics area, and I did see psychokinesis experiments first hand from two people. Believe me, it's very shattering to one's conventional understanding. I sat in a large dining-room and a lady came in and sat down beside me. It was Nelya Kulagina and she took off her gold wedding ring, set it on the table about four feet in front of her, sat back, folded her hands quietly in her lap and moved her head just slightly and that ring began to move across that four feet in jerky steps. It covered the four feet within a time period, I would say, of the order of a minute.
>
> She then took a plastic pen top from one of the members and set it vertically upright on the table, again in the order of three or four feet away from her. What happened was the pen top moved across the table staying vertically upright. So that was rather impressive.

Apart from moving objects using only the mind — Tiller gave many examples — he then went on to link such effects to energy pathways, measurable through linking voltages between acupuncture points, such that healing processes could be harnessed.

> With my studies, using a biomechanical transducer — which is just a wand in a hand of a person who can move energy like

a dowser — energy flows into the individual through the left hand, through the trunk of the body and to the right hand.

This is one of the three main energy circuits in the body and one way that this has been utilised in my own life is healing myself from a lower back problem. I find that what I can do is to place my hands on my back — I do it each morning — I place them as high up as I can reach, for about two minutes, and then just move it down a palm width for about two minutes, and down lower for about two minutes until I get right down to the coccyx area for about two minutes. The whole thing takes in the order of ten minutes and I can feel warmth being generated in the back and muscles relaxing.

Tiller went on for nearly the entire program. At the end I asked a question about this 'energy' and why it should help us:

RW: Do you find any of these ideas in conflict with what is your basic training as a scientist, physicist, and as a professor of materials and engineering at Stanford?
Tiller: I don't find any conflict to myself. My colleagues may find it a bit of a conflict, because to me it is science. It is, to me, tomorrow's science and it's going to take a great deal of work to make it as reliable and as rigorous a science as our presently accepted science, but nonetheless, I think that's there in the future.

And finally, his boast. You don't get too many scientists, let alone engineers, coming out with this kind of triumphal guff anymore:

My feeling is that we are on the threshold of a revolution — a revolution so vast, much more vast than this world has ever seen, even in the days of the Greeks. We are talking about a revolution of scientific understanding.

Professor Tiller is still with us, though he retired from Stanford in 1998 with full honours for his conventional work on materials. I quote him at length because *The Science Show* in which he starred — only our sixth — proved to be the first spectacular ratings hit. The letters arrived in big boxes every day, asking for transcripts. Something had erupted in the listening population — and that population turned out to be huge: 50,000-plus in Melbourne alone, before any repeats.

Remember, this was the Age of Flake, but it was underpinned by some respectable writing. One of the foremost exponents was Ivan Illich, author of *Medical Nemesis*, published in 1975, just in time for our new program. Illich criticised medical hierarchies as reminiscent of the Vatican, with whom he was at odds. He also attacked educational systems that devalued intellectual rigour and human interaction. Illich graced *The Science Show* many times and was immensely popular.

> In a society the environment is rearranged by a professional group which provides health care for people. The healthcare system does this by extending a medical, professional monopoly over what people need when they feel sick.
>
> Medicalisation of society creates a myth that the management of my health problems by an outside agency is more valuable than doing it on my own.

Illich had a few valid points, of course. It is the brusque, impersonal encounter with many a harried doctor that sends some citizens across to 'complementary' medics where they may be heard at greater length and taken seriously. The amount Australians spend on this kind of medicine remains enormous, close to the expenditure on orthodox therapies. And, in our broadcasts back then, we had learned to push the populist button. It is rather easy, I must say. Select some favourite contrarian topics: diets, healing, vaccination or fluoridation, ESP, dreams, how horrid the professionals are, and allow the all-too-willing pronouncers in the field free rein, and you can have ratings figures out of this world. The temptations are still there today and are yielded to disgracefully in Australian media. Meanwhile, they, in turn, call ABC and SBS 'elites' in some sort of pejorative sense, implying we are out of touch with real people and what they like and value.

Gradually, for me, the intellectual justifications began to fade. I had been influenced by ABC colleagues such as Malcolm Long (still a close friend) and Peter Fry whose book *Beyond The Mechanical Mind* had explored these anti-establishment themes. Add R.D. Laing, Illich and various other sceptical authors and you had a good, credible justification for pushing the boundaries.

But two things happened, in the end. The first was a ludicrous interview I did with a couple from the Southern Highlands of New South Wales claiming to be experts in 'past lives therapy'. I must say I did this originally as a laugh. The woman said she lived, in a former life, on some far distant planet running a space station. She was in charge of Star Wars Central! 'What is it like?' I asked. 'Describe it.'

All she could say was that it was big. 'Big,' she exclaimed for the fourth time. I closed off the interview forthwith and swore never to waste my time again, even for a frolic.

Then, shortly after, obviously fed up with all the Uri Geller bullshit, Von Däniken ravings about aliens coming to Earth, ufology and dowsing balderdash raining from every cloud, Phillip Adams and Dick Smith got together to form the Australian Sceptics, of which I am a proud member. Other offshoot outfits such as Friends of Science in Medicine are also doing fine work today, having faced similar outrageous provocations — vaccines, for example!

At the height of the Age of Flake even *New Scientist* magazine put magician Uri Geller's photo on its cover, asking whether his claims to supernatural powers were real.

Maybe that was a final clincher for me. As well, we were about to hit the very different 1980s. The Great James Randi,

Geller's nemesis, came to Australia and we recorded an interview, during which Randi bent some spoons for me — before my very eyes! And I still don't know how he did it. But Randi showed how good it is for scientists, every now and then, to get a little help from show business. Scientifically sound show business, of course.

> If we were more sceptical about the promises that lawyers and politicians and the aforementioned creature make to us, I think we would be far better off. It's the gullible people, the people who aren't sceptical, who fall for all kinds of scams and schemes. And the scoundrels are out there to get them, and they will get them, every time.
>
> Just because you don't have an immediate explanation, do not assume that the answer is paranormal. It is within the confines of science to explain all these things. We do not need to invoke the supernormal. There may be such a thing but in thirty-five years of looking for it, I have found nothing.

2 Enter Norman and Crunch Time
The end of the seventies frolic

He was bearded and a bit tubby. He seemed to have about five jobs at once: consulting about management, writing for *The National Times*, practising as a paediatrician somewhere on the distant outskirts of Sydney, giving talks. And there was something else — was it spying for Scotland, undermining the Israeli government or preparing a list of Australia's favourite doctors to demolish? He was a thirty-six-hour-a-day man, even then.

Norman Swan hasn't changed much. He's slimmer now and there is no beard, but his jobs amount to dozens and he runs what we call Swan Industries, whatever they may be. He doesn't disclose much, but they include a TV series (*Tonic*), occasional treatment for friends and emergency consultations (he undoubtedly saved the lives of several colleagues — myself and Mark Colvin included), and ... well, it beats me — maybe he's still spying for Scotland.

Who else would fly to Glasgow for three hours, then turn around and fly straight back? He claims it was to take his aged parents out for an anniversary tea. Sure!

Norman did a few medical items for *The Science Show*, then a four-part special on Israeli politics for the equivalent of *Background Briefing* and then invented *The Health Report*.

I suspect all that was in one week. We have unusual work proclivities in the ABC Radio Science Unit.

His arrival meant I could do less on medicine, which was a relief. The topic is of huge interest to the public but it is incredibly labour-intensive. And one quality Norman brought to us was pitiless questioning. Even Nobel laureates quaked when they saw him in the front row of a press conference, knowing they would risk being shredded by Norman's vast knowledge and relentless force of inquiry. (The fact that many, including me, find his Scottish accent hard to fathom is another hazard. He actually rhymes Perth with *TERR*-orist, instead of saying *PURRR*-th as any gentleman would. And as for that word necess-*AAARILY*...!)

He has now done *The Health Report* for more than thirty years and it is a triumph of rigour and range, and has a huge professional as well as a lay audience. Norman has also, several times, done *Science Show* specials on various notables whose professional practice perhaps deserved scrutiny. The first and most spectacular was to accuse Dr William McBride, society favourite and whistleblower on thalidomide, of scientific fraud. McBride never recovered. Even now, fifty years after the thalidomide story broke in the *Sunday Times* of London, some of those (Australian) journalists there are uneasy about our exposing of Dr McBride's role.

But Norman was just the finely focussed mind we needed as the silly seventies turned into the unforgiving eighties.

* * *

The 1980s arrived with a crunch. The green option was not a cosy love-in with multi-coloured banners, it was war. The first time I attended a meeting of the Australian Conservation Foundation in the mid-1970s, it was held at the Academy of Science and chaired by Prince Philip. Not a brickbat in sight. Now we had the Franklin River, with fighting, arrests — and the world was watching.

Meanwhile, in 1981, the first known carriers of AIDS had reached North America. The germ, HIV, would be revealed only two years later, in 1983. A new plague to frighten everyone. Norman had arrived just in time to enable us to keep up with what was being done scientifically. He provided many *Science Show* reports on the crisis.

Fraud in research was also being revealed in several scientific locations. Some fraudulent claims had inadvertently been broadcast on *The Science Show*, it would turn out. Dr Michael Briggs, foundation Dean of Science at Deakin University, had been embroidering his research findings on the contraceptive pill. It is likely that the subsequent scandal and inquiry killed him. American science journalist Nicholas Wade wrote a book about research fraud, called *Betrayers of The Truth*, in which he offered the observation that it was often very clever men (like McBride) who cut corners with lab work, because they thought the conclusions were obvious and the research tedious, so they'd just skip to the findings without wasting time!

The seventies seemed transformed. But we still allowed a certain breadth to our coverage of science. It made sense that dancing and music may be linked to dyslexia, as one of our guests averred, and she showed that doing both would help afflicted children. We also had the American writer Jory Graham talk about dealing with death. She was herself given limited time to live and talked to us with humanity and insight. Then the Feingold diet made an impact: it was a scientific application of elimination diets, especially of colourings and flavourings, and could help children recover from hyperactivity, now better known as one of those cursed acronyms: ADHD. The response from the audience was phenomenal. I was invited to talk to conferences of nutritionists.

Again, we were not really aware of the change in the zeitgeist until well into the new era. And there was plenty of diverting research to keep us busy.

Enter Norman and Crunch Time

* * *

Prince Charles opened the Anglo–Australian Telescope (AAT) at Siding Spring Mountain in 1973. This gigantic 'light bucket' would fulfil its promise quickly as one of the best instruments of a new generation. Its site, near Coonabarabran in New South Wales, was also very lovely and ancient. Animals were everywhere. When making a documentary about the telescope, I recorded local currawongs and then had a composer write me music. Such were the luxuries of yesteryear. That program went to thirty-two nations around the Commonwealth.

Why such a vast investment in our small nation? Why, for that matter, all those telescopes from Parkes, New South Wales, to Hobart (where radio astronomy was pioneered) to Western Australia and beyond? The answer lies in the map. We did have a small population but the island continent offered a terrific view of the southern sky and, before the AAT, little of it had been explored. Above us, views of those two small galaxies — the Magellanic Clouds — and the larger Milky Way are substantially different as seen from Australia. OK!

One of my first interviews about all this was with Professor Fred Hoyle who chaired the board setting up the new telescope on Siding Spring Mountain. Now, this book will be replete with names of past notables — unavoidably so — but do not neglect the name Hoyle, if you've never heard of him. He was an astrophysicist of the highest rank and should have received a Nobel Prize for working out where the elements come from. They are cooked in stars. When the old star explodes it sends all those metallic and other elements out into space, eventually, perhaps, to be concentrated as part of planets. Then, on Earth, at least, comes life. So *you* are truly made of star stuff.

But Hoyle misbehaved. He was difficult in the hierarchies of science; he was not a compliant committee man — he had too much residual northern chippiness. What's more, he even became a relatively successful novelist; but he also came up with wild ideas. It was he who proposed that AIDS and other plagues came from the sky, having been formed somewhere in the universe. Hoyle would refer to patterns of infection that matched celestial weather and claim that such patterns bore him out. Flake again! He also doubted Darwinian evolution. One of the reasons he put germs in space was his inference that it would take too long for them to be shaped on Earth. It was he who espoused the nonsense that forming a living thing on this young planet would be like a wind hurtling through a junkyard and forming a jumbo jet. So, no Nobel.

> I don't believe that when one looks at the mathematics of finding these exceedingly complex substances that there is any chance of doing this by random shufflings.
>
> Life is really based on the commonest molecules and the amount in our galaxy alone is huge, the amount when our sun formed was simply vast compared to this thin skin of stuff that we have on the surface of Earth. I wouldn't think that life originated in our own system; I still think that the processes that occurred were simply amplifying an already existing life form.
>
> What happened in the early days is that if you take a bacterium, keep it in the correct conditions, nutrients, and so forth, it will multiply and in a week its mass will be equal to the whole Earth, in a couple of weeks the mass of the resulting progeny is equal to the whole galaxy and in three weeks it's equal to the universe. You know that multiplying business. So it's just a question of getting it right. One only needs a few viable things and the whole thing explodes.
>
> I think that's what happened in the early days of the solar system. And it happens when many stars are formed; it's happening all the time.

It is really staggering that such a sophisticated physicist should be so clumsy intellectually as a biologist. Perhaps the fiction writer had invaded the sceptical part of Hoyle's otherwise fabulous brain. I found him personable but dry and looked him up after his retirement. He was always willing to talk. Latterly he was in the Lake District in the United Kingdom where he relished long walks through the hills. When his wife became ill he moved to Bournemouth where I suspect he felt marooned in the tea-cosy niceness, washed up. Some stars just fade away.

Another Cambridge colossus is Stephen Hawking. I expected him to fade fast too, but here he is (at time of writing in 2015) doing nights at the Sydney Opera House as a hologram, 'live' from UK and selling out all tickets in fifteen minutes! What is it about Hawking?

I've interviewed him twice, both times at his old office off Silver Street in front of his large portrait of Marilyn Monroe. He was friendly, jokey and good to be with, apart from that initial embarrassment one feels after having asked a question, to then have to wait two minutes as he prepares an answer on his machine. Do you look around the room in silence — treat him as if he's not there? Or do you chat at the risk of distracting him? On the second visit I allowed myself brief comments. He seemed to appreciate that. In my day, he had some small hand control left, now he must twitch a cheek muscle to operate the computer.

We talked about his views on the universe, well known now, and that old furphy about few readers finishing his book, *A Brief History of Time*. I got our Shakespearean star John Bell to read Hawking's considered reply to this (Bell was chosen as another joke), for the late astrophysicist John Bell was one of Hawking's chums. This is his answer to the critics:

> I first had the idea in 1982 of writing a popular book about the universe. The intention was, partly, to pay my daughter's school fees, although by the time the book actually appeared she was in her final year of school. But the main reason was I wanted to explain how far I felt we'd come in our understanding of the universe. How we might be near

finding a complete theory that would describe the universe and everything in it.

If I were going to spend the time and effort to write a book, I wanted it to get to as many people as possible. I contacted a literary agent, Al Zuckerman, who had been introduced to me as a brother-in-law of a colleague. I gave him the draft of the first chapter and explained I wanted it to be the sort of book that would sell on airport bookstalls. He told me that there was no chance of that. It might sell well to academics and students but a book like that could not break into Jeffrey Archer territory.

I gave Zuckerman a first draft of the book in 1984. He sent it to several publishers, including a fairly upmarket United States book company. But I decided instead to take an offer from Bantam, a publisher more oriented to a popular market. In fact, I think my book was the first science book that Bantam had accepted.

I had been very impressed with Jacob Bronowski's television series, *The Ascent of Man.* Such a sexist title would not be allowed today! He gave a feeling for the achievement of the human race in developing from primitive savages only 15,000 years ago to our present state. I wanted to convey a similar feeling for our progress towards a complete understanding of the laws that govern the universe. I was sure that almost everyone was interested in how the universe operates but most people cannot follow mathematical equations. I don't care much for equations myself.

Even if one avoids mathematics, some of the ideas are unfamiliar and difficult to explain. This posed a problem. Should I try to explain them and risk people being confused? Or should I gloss over the difficulties?

There were two such concepts in particular that I felt I had to include. One was the so-called 'sum over histories'. This is the idea that there is not just a single history for the universe. The other idea is 'imaginary time'. With hindsight,

Enter Norman and Crunch Time

I now feel that I should have put more effort into explaining these two very difficult concepts. Particularly 'imaginary time', which seems to be the thing in the book with which people have the most trouble. However, it is not really necessary to understand exactly what 'imaginary time' is, just that it is different to what we call 'real time'.

Time magazine published a profile of me. This encouraged Bantam to make the print order rather large; even so, the editors were taken by surprise by the demand.

Why did so many people buy it?

In the proof stage I nearly cut the last sentence in the book, which was that 'we would know the mind of God'. Had I done so, the sales might have been halved.

My wife was horrified, but I was rather flattered to have my book compared to *Zen and the Art of Motorcycle Maintenance*. I hope, like *Zen*, that it gives people the feeling that they need not be cut-off from the great intellectual and philosophical questions.

It's been suggested that people buy the book because they've read reviews of it or because it's on the bestseller list but they don't read it. They just have it in the bookcase or on the coffee table thereby getting the credit for having it without taking the effort of having to understand it. I'm sure this happens but I don't know that it's any more so than for other serious books including the Bible and Shakespeare. On the other hand, I know that some people at least must have read it because each day I get a pile of letters about my book, many asking questions or making detailed comments that indicate they have read the book even if they don't understand all of it. I also get stopped by strangers in the street who tell me how much they enjoyed it.

The frequency with which I receive such public congratulations, to the great embarrassment of my nine-year-old son, seems to indicate that at least a proportion of those who buy the book actually do read it.

> My agent suggested that I allow a film to be made about my life but neither I nor any of my family would have any self-respect left if we let ourselves be portrayed by actors. The same would be true to a lesser extent if I allowed and helped someone to write my life story. I try to put people off by saying I am considering writing my autobiography. Maybe I will, but I'm in no hurry. I have a lot of science that I want to do first.

Well, he did write his life story. And, as we now know, self-respect was put aside, perhaps, as we now do have a film of Hawking's life, based on his ex-wife Jane's book. But I find it, as you do, almost beyond belief that Hawking has lasted so long at the top of his powers. But for me the greatest delight was in the final Monty Python live stage show. He had a small part in the filmed insert cavorting along The Backs in Cambridge in his wheelchair. But afterwards, as we returned to the live performance, there he was, in a box in the theatre itself, with Lucy his daughter triumphantly holding Stephen's arm aloft in greeting. Talk about cheerful resilience!

* * *

Let's stay with astronomy, which is always popular with audiences, though I often wonder how much of the technicalities and boggling physics the public take in (I have difficulties too). Scandals are more rare in this area, but some do arise.

The Parkes Radio Telescope is as famous as any dish we have. It carried the Apollo 11 moon walk and helped track the stricken Apollo 13. It is also responsible for finding about half the pulsars in space and plenty of quasars too. Those pulsars had themselves been discovered at Cambridge by student Jocelyn Bell. She reported to her professors that she had noticed a precisely regular signal from deep space. The dons promptly decided to keep it secret for a while in case it was from ET.

Then they worked out that the pulsars were ultra-dense bodies revolving at high speed. The professors published and got the Nobel Prize. Jocelyn Bell Burnell is to this day phlegmatic about the setback.

> We spent about two months trying to convince ourselves, first of all, that the equipment was behaving properly. That was the scary bit for me because I had built the equipment. I was dead scared that I had done something wrong and all these brilliant Cambridge brains were going to discover what a fool I was. That would be the end of my doctorate. But the equipment was OK and then we started wondering whether we were picking up interference of some sort because we were looking for very, very sensitive cosmic signals. And the locally generated things, like sparky thermostats, arc welders, these kind of things, all produce radio radiation, and in sufficient quantity, to swamp the cosmic stuff.
>
> I can remember one evening just before Christmas, the evening before I went home, having a meeting and saying, look, we don't really think these are signals from little green men, but we don't have an alternative natural explanation. How do we publish this? We didn't resolve the issue that evening.
>
> I was in later that night, doing more of this chart analysis, and discovered what looked like a second source of these curious blips without the telescope. At the dead of night I got the telescope to work and found that there was a second one of these sources. And that really was sweet, that's the sort of Eureka moment. Discovering the first one was a worry, discovering the second one a couple of months later was absolutely great.
>
> **RW:** [Radio astronomers] Antony Hewish and Martin Ryle got Nobel Prizes. You didn't. You are one of the people who is often named as a neglected woman in the laureate stakes. What do you think about that?

> **Jocelyn Bell Burnell:** I think there's a tremendous sympathy vote and you can do very well out of not getting a Nobel Prize — witness me being here in Australia on this lecture tour, which is great, absolutely superb. So I'm doing very well, thank you.

Nowadays the serene Professor Dame Jocelyn Bell Burnell (she was educated at a Quaker School) is professor of astrophysics at Oxford. She has presented poetry on space at many a sublime evening.

Earlier on, at Parkes, I had spoken to its way-out English director, John Bolton. He was an enthralling, well-adapted new Aussie, with all the slang, rolled up fags and diverting indiscretions you rarely come across in top astronomy. The locals loved him. One of the tales he told with one leg up on a pole fence, a ciggy dangling from thin lips, was about why Apollo 13 nearly died. He told me they had tried to lift the space module from the ground in the hangar, without having remembered to unbolt several of its parts connected to the ground. The vehicle cracked in several places. Most of them were repaired — apart from the temperature controls that caused the blow-up in space. Fortunately, as the *Apollo 13* film starring Tom Hanks shows, a few computer geniuses and engineers at NASA saved the day.

Bolton was a top astronomer, now largely forgotten, except among the professionals, and is always enthusing about the work Parkes did, and is doing:

> The most distant radio galaxies and the quasars are emitting energy on an incredible scale compared with that of our own galaxy. The fact that we can see them and detect radio waves from them must mean that they are intrinsically very luminous and physical processes of degeneration are going on in them of which we have no concept here on Earth.

We had some superb instruments, internationally, as everybody knows, though the beginnings have often been as dicey as with Apollo 13. The Hubble Telescope was launched into space only to find that the wrong dimensions had been used to set its vision. Expeditions went up there to fix everything, and it worked: the skewed gaze was fixed with special corrections, like putting new glasses on a myopic cosmic granny. Now, after a triumphal career, Hubble has lasted ten years beyond its expected span. The same with shots to Mars — so many crashed or failed. I asked astronomers at Tucson Arizona how they coped with years of meticulous preparation only to have their treasure smash in seconds. They said quietly: 'That's the deal; you know what to expect. Space is never easy.'

Nor is the physics. We now have a cosmic embarrassment with the two great models unable to mesh: relativity and quantum mechanics. Add the mysteries of the two 'darknesses' — dark matter and dark energy — and it is obvious that we are poised to see a revolution in the next few years. We have the newly powerful Large Hadron Collider in Switzerland poised to find new particles following its success with the Higgs boson; we have the gigantic new telescopes like the Giant Magellan Telescope here on Earth and the James Webb due up in space. The James Webb will have an eighteen-segment, 6.5 metre mirror and operate 1.5 million kilometres from Earth after its launch in 2018. Then there is the SKA.

The Square Kilometre Array (SKA) in Western Australia and southern Africa will explore regions of the cosmos never seen before. Dealing with the information SKA collects will require a computer with the processing power of about one hundred million PCs. Welcome to Big Data.

Put all this technical power together, combined with a generation of super-smart astrophysicists (lots of them women), and you can guarantee huge revelations. And the public is immensely interested. In the UK, when CERN started the new collider, it was the number one item on the national news —

even before finding the Higgs! Why? Because, as I said in the beginning: THIS IS REAL. It is also important.

Over the forty years of *The Science Show* the lesson has been that doubts about such things as the existence of black holes, as there was in the beginning, can be swept away as separate lines of research all begin to point in the one direction. As with climate change, Darwinian evolution and biodiversity, so many distinct pieces of evidence coalesce almost to a certainty. Yes, one needs always to adjust aspects of the science, but there comes a point when an edifice is clear. We have an answer.

This should happen in cosmology and other aspects of physics quite soon.

How to keep up with the adventure? In two ways. First, connect to the action. There are several lines of astronomy such as Galaxy Zoo, a program operated out of Oxford, in which thousands of volunteers classified almost a million galaxies thus saving the boffins years of work. Otherwise, continue to listen to *The Science Show,* watch *Catalyst* and tune into *Star Stuff,* presented on-line by Stuart Gary, and you'll be briefed.

The other way is to ask your MP to help foster Australian space research. Brian Schmidt, our recent Nobel laureate from Mount Stromlo Observatory and now ANU vice-chancellor, has noted just how far behind we are in Australia, notwithstanding the SKA. His warnings are stark.

> **RW:** An awful lot of people tell me that with China producing something like a million graduates in science a year and 400,000 graduates in engineering a year, pretty soon, at that rate, we're going to be bystanders.
> **Brian Schmidt:** What happens if you don't do these things? You don't get new ideas, you keep on doing things you already know how to do, which means that you make a better toaster, but you don't come up with a new idea of a microwave oven. And so if you want revolution you've got to let people go through and do things that they don't know the

answer to, and that's the beauty of it. But I believe we've sort of slowed down and we need to get the foot on the accelerator again.

We've come a long way since the AAT, now forty-two, was the shining new emblem in Siding Spring Mountain. And so did our astronomy, led by Brian Schmidt, whose Nobel Prize came from studies of supernovae, begun in Harvard, and which showed that the universe is expanding at a much faster rate than was thought possible. Blame dark energy, whatever that is. It may mean that one day, telescopes or not, we may be too far from other galaxies to see anything like the cosmos we do now. A lonely planet? I suspect we won't be here to tell.

But let the last words on this section go to someone speaking on the 14 July 2015, as the *New Horizons* probe, after a voyage of five billion kilometres sped past Pluto. Everything worked, as it had with *Curiosity* on Mars and the Rosetta mission, landing its vehicle on a comet. The scientists met all this triumph, not with smugness but with emotion. One woman, the Mission Operations Officer, stood out:

> **Alice Bowman:** I can't express how I'm feeling to have achieved a childhood dream of space exploration. I'm pretty overwhelmed at this moment and I just want to say thank you to everyone, and please tell your children and anybody out there: do what you're passionate about. Don't do something because it's easy, do something because you want to do it. Give yourself that challenge and you will not be sorry for it.

So here we go, out from the solar system.

* * *

And then there was AIDS. Just as the 1970s and 1980s forced astronomers to face new realities after the romance of Apollo, so medical research lost its avuncular calm. This was not so much the kind of organisational challenge Ivan Illich had thrown at the doctor–patient relationship (though an element of that did intrude) but more one of scientific understanding. What was this blight killing young men? Out of nowhere, it seemed. But now signs are clear going back long before the 1980s, that a terrible disease was going to emerge if it infected carriers who then travelled far. The rest we know. For science journalists it was difficult. The general media often made the task worse. Dealing with what some media meatheads called the 'Gay Plague' was tricky and daunting. In the beginning we talked to doctors in the field about their patients and how bad the infection and death rate was becoming.

> **RW:** AIDS — Acquired Immune Deficiency Syndrome — is a modern plague. You've heard about it countless times but four or five years ago practically no one, outside a few scientific circles, could tell you what it was. Now there's a new book, *AIDS and Australia*, to tell you what you should know and the authors are Dr Julian Gold, director of the newly opened AIDS clinic in Sydney, and Dr Alistair Brass, who edits the *Medical Journal of Australia*.
> **Alistair Brass:** The virus does seem to have changed a bit. Certainly the pattern of infection that they're discovering has existed in Africa for some time is a little different from the pattern we're seeing in America and now in Australia and Europe, so there is some suggestion that the virus may in fact, in addition to being a rather complicated and unusual virus, also mutate and modify itself rather as the influenza viruses do over the years, which makes it even more difficult to pin down.
> It's an unusual virus in that the protective coat it generates for itself seems to be much thicker and more resistant than most viruses. This means that even if you do have antibodies floating around, which is what a vaccine

would do — a vaccine, if you gave it to people, would promote a healthy level of antibodies to the virus — it may be much more resistant to antibodies than our other viruses such as the flu virus. One of the things they seem to have found in unravelling the sequencing is the code for the coat. This, perhaps, does not change so much as the rest of the virus as it modifies and mutates and if they can really pin that down and they find the coat stays the same as the rest of the virus changes, they might be able to make a vaccine which could then, perhaps, generate some antibodies, which could then, perhaps, destroy the virus coating. But all this is very speculative and probably several years down the line.

He himself died not long after the interview went to air — of AIDS. The commentaries and warning videos on AIDS were uncompromising. Who can forget the bowling-alley version showing us all being knocked down like helpless nine-pins?

In fact, the science moved quickly and well. After a small skirmish over priorities, French virologist Luc Montagnier identified HIV as the cause of AIDS and got the Nobel Prize. Robert Gallo of the USA was close, but seemed not to mind being passed over. Although there is no cure for AIDS, drugs are now available to stabilise the patient so that a shortened life need not be the inevitable curse of yesteryear.

All this happened over twenty-five years — not a long span for a science project. Very fast work, in fact. Soon perhaps, as Professor David Baltimore has often told me, a vaccine may be forthcoming. We hope so. Professor Baltimore is a former President of Caltech, and of the American Association for the Advancement of Science (AAAS). He won the Nobel Prize at thirty-seven for working out the mechanisms behind the resilience of the AIDS virus. In fact, one of my first interviews with him was on his being awarded the Nobel Prize forty years ago. This is what he said about AIDS recently.

> Twenty-five years after we discovered HIV, we're still saying [the vaccine] is at least ten years off. The big hope that we had was to make a vaccine that would induce T-cell immunity and protect people through T-cells. I mean there are lots of reasons to say that should not work but we were hopeful that it would work because it is the only thing that we really have. Those hopes were dashed this year by the Merck trials that failed and were stopped prematurely because they were failing so miserably. In fact, there's non-statistical but suggestive evidence that the vaccine that they had made it easier to get HIV infection rather than harder. The one thing we can be sure about is that the vaccine did not do any good. So now everybody is back to the drawing board. There's a huge debate in the community about what's worth testing next — is anything worth testing next that we have around today? There are a lot of people scurrying around saying we've got to start thinking in some different direction if we're ever going to produce an HIV vaccine, and maybe we never will.

Brutally honest, as always.

3 Here Come the Women

A scientific revolution

By the mid-1980s, the Franklin River was safe once more — the Labor Party had made the proposed dam an election issue and the nation voted for a river unimpeded.

Soon, our new science minister, the formidable Barry Jones, was building his message in his book *Sleepers, Wake!* into his speeches. Was the nation listening?

Sleepers, Wake! was a recipe for what to do when Donald Horne's 'Lucky Country' ran out of luck (or commodities, as the message had it). We had to seize the times, note the new communication technologies just being born, and become part of the revolution. Jobs would require new skills; international alliances would need to be made and refreshed. The rewards would be immense. While Jones spoke, more holes were dug, more sheep sheared. Little changed.

Australia is promoted, not least by itself, as a bold country, willing to embrace change, say boo to the boss and be creative. Donald Horne was sceptical — and so am I. So much was achieved *despite* pathetic leadership and unending blockages. Fortunately, Australia is a big country and there are spaces where the prominent poltroons are not looking to interfere and where one can make things happen. Or go overseas, as the previous generation of scientists had done.

It is worth looking at Tom Barlow's books on these myths. I first met Tom in the Senior Common Room at Balliol College in Oxford. The Master of Balliol sidled up to me and whispered, 'There's an Aussie at the end of the table who's just written a film script about a taxi driver. And he also writes for the *Financial Times* about science policy.' The film turned out to be an embarrassment, but Tom has since published several books on innovation and been an adviser to an Australian Science Minister, Brendan Nelson.

Tom Barlow: Australians have always been terrific at scavenging other people's technologies. You still can't scavenge as a highly innovative business unless you're doing the science and technology yourself. And secondly, and most importantly, you won't generate the bright sparks who can understand the science and technology elsewhere in the world unless you've got real frontier science happening in your universities and public research agencies.

RW: There's a tendency in Australia to say 'Oh, we're world class' and I always found this rather cute because it seemed to be a desperate thing to say, 'Look at us, look at us, we're good after all'.

TB: Like you, I think it's hilarious. There's another phrase as well, which is 'Australians punch above their weight'. I mean especially since the weight of Australians is steadily increasing, we are in the middle of an obesity epidemic. But look, they are both completely silly and consistent with a kind of vanity actually that a lot of Australians have about our national ability to have ideas. If you look at the facts, though, Australians publish just 2 per cent of the world's scientific literature, which is reasonable compared with the size of our population, which is 0.3 per cent of the world. But entirely consistent with the size of our economy, which is around 2 per cent of the economies of the developed world, where most world science occurs. And, in fact, on per-capita terms even, we're not that impressive. If you look at Switzerland, it produces almost twice as many scientific articles per person as Australia does. So we boast about these things but we're not that great.

You look at patents and we are even less impressive. We produce around 1 per cent of the world's patents. There are good reasons for that, to do with our industry structure, and our strengths in industry are often in areas where people protect their intellectual property through other means. In some ways the hilarious thing about the phrase is not so much its accuracy but actually the meaning of it. I mean, what does it mean to be 'world class'? It sort of means you're in a class that's of world standing. Actually, we should be aspiring to be the best in the world.

* * *

More on innovation and what we could and should be doing, later. It will be a constant, bleak refrain as we move through the decades. First, I want to highlight a real change during those forty years: the arrival of women in force. Yes, they had been there in ones or twos — more of a curiosity in previous times — but now the difference was becoming unmistakable, though the change was far from smooth. Even now there is a paucity of female talent at the very top, despite enormous changes at junior and middle levels. Is the glass ceiling really made of tungsten?

One of the new Whitlam government's innovations in 1973 was an adviser on women's affairs, Liz Reid. Many were surprised at the flack she took and how, sometimes, she seemed to be overwhelmed by the hostility and opprobrium. What, I wondered, was making so many men so cross? Then it began to dawn on me, something Germaine Greer had said that I had dismissed as rhetoric: that lots of men *hate* women. But why?

Since then, a number of reasons have become obvious. The first is the agony of male adolescence. Girls, who counted for very little when young, suddenly become transformed in male eyes into impossibly alluring creatures. They are quickly snapped up by young men, leaving the adolescents grieving and deeply frustrated. I remember feeling just this anger and loss up to the age of about seventeen. Then I grew up. I realised that most attractive women feel unattractive, blemished, too large, vulnerable, and nowhere near like the supra-confident Bond Girl with two brains immune to all suggestion of flaws. It is good to grow up. Many men don't. Their brains aren't mature until about twenty-seven and even then they are not necessarily smart socially.

Secondly, there is the male reaction to whatever workplace emancipation women have achieved. Lots of blokes resent it and undermine women on the sly, or even blatantly. It is the same mechanism you observe among many Americans who loathe the social changes they have seen since the 1960s and who want to turn the clock back to when Mum and Dad looked like the icons in the painting by Norman Rockwell. Now we have smart women 'leaning in' and I promise you, a lot of chaps want them to stop.

Thirdly, there is what Geraldine Brooks has called women's 'Nine Parts of Desire'. (That's out of ten!) Men are way down in this capacity — say four or five. Brooks was referring to an Islamic assessment of female sexual passion and the consequent need to put a stopper on it. But can it really be true that the capacity for sexual enjoyment of women is so very high: nine out of ten is vast, full of multiple orgasms and the temptation to stray, surely?

Well, in 2005 (so recent!), Dr Helen O'Connell, a urologist based in Melbourne, dissected the clitoris to find it to be much bigger than anyone dreamt: not a little button but an enclosing web as large as the tissue in a mature penis. An organ on such a scale is obviously much more likely to enhance stimulation and satisfaction than the 'button'. No wonder possessive males have done all they could to diminish the response. Some cultures even developed genital mutilation for young girls, which must be one of the most repulsive practices still so ubiquitous.

Why would women need to feel so turned on, if allowed? Well, just imagine there is a sweaty, lumpen male, usually with all the subtlety and grace of an inebriated diprotodont, and he wants to stick his malodorous willy on the way to your womb. Why would you *ever* want to let him, unless there's a payoff?

That O'Connell's discovery had to wait to the twenty-first century is both mystifying and a disgrace.

RW: Many diagrams of a scientific, and even a medical nature, just draw the clitoris as the gland. As if it's a rose bud with one nerve, like a long wire attached and that's the end of it.
Helen O'Connell: Certainly, if that's your concept of what this is, it's entirely wrong and it's probably much more than ten times wrong. And then there're the anatomical texts and I would think that the mass of tissue that we have typically found is probably about double what's in the good anatomical texts such as *Grey's Anatomy*.

The reason things are not depicted as you would expect them to be, given the knowledge, is very hard to work out. I guess science doesn't work in a vacuum; it works as part of society in general and so you just assume that all this stuff is right. You can see that there's part of the inside of the prostate that's said to be analogous to the uterus, for example. It's a minute little thing and so presumably it was thought that the clitoris was minute because it just is.
RW: Does it tell you anything more about women's sexuality?
HC: Yes, I think it does. We haven't done any functional work at all but there's certainly quite a bit of erectile tissue on the lateral walls of the vagina nearly completely surrounding the urethra. It's probably got quite a lot of potential for engorgement, stuff that we weren't aware of, probably intuitively all women are aware of it but then you don't have the pictorial basis for it.

> A relationship between this work and sexuality in general is obviously very complex and it's an area where there's so much lack of understanding.

You would imagine men would be keen to enhance a woman's sexual enjoyment, to make her more ready and willing — but not if you are self-absorbed and concerned only about possessing. Someone else's feelings are then immaterial. And too flamboyant a female response would intimidate many men.

So much for anatomy. Intellectually, women were also cut off. Did you know that it was Ada Lovelace who, in 1842, dreamt up the idea of an internet, machines that can represent concepts in maths and interconnect? What we now call laptops and clouds. She had been computer inventor Charles Babbage's offsider in designing his machine, but she was also the world's first computer programmer. Better known as Lord Byron's daughter, she is beginning to be recognised. Bryan Appleyard, in the London *Sunday Times*, called Lovelace's 20,000-word summary of the potential of computers (Babbage's 'difference engines') one of the greatest pieces of writing in the nineteenth century.

Emilie du Chatelet is another of my heroines. Fortuitously, she married a general, which afforded her lots of time to have affairs (with Voltaire, among many) and other fun while her husband was away fighting. Jealous suitors were kept away by her superb swordplay. In her spare time she read physics and did maths. When she knew she was going to die as a result of giving birth (in her early forties!), she spent her last weeks furiously translating Isaac Newton's physics into French and *correcting* some of his infelicities! It was a real tour de force.

There are many more extraordinary female achievers. A final example: Beatrix Potter.

(Mycologist) Roy Watling: She was experimenting with the development of spores of fungi. In about 1895 she got a microscope and a little drop of water and put the spores under the microscope and watched them germinate. And as they produced their little germinating hyphae, she drew them and then drew them again an hour or half an hour later and then half an hour later after that. So she had these illustrations covering the development of these spores into a single hypha, and then into a mass of hyphae.

And believe it or not, she found that some of the mushrooms that we find in the woods and fields, they had asexual stages just like moulds, like penicillin and other types of microscopic fungi. And that was a first.

Sharon Carleton: Apart from being an interesting historical anecdote about a famous author, would it have made any difference if her paper [presented at the Linnaean Society of London in 1897] had been published?

Roy Watling: It would have allowed people to say she was one of the first British people to germinate and study these developing spores as opposed to having all these years of conjecture.

Potter is one of many women whose accomplishments now come as a surprise: a pioneering botanist as well as a supreme artist and delightful storyteller for children. Later in life she helped establish what would become the Lake District National Park by purchasing all the sheep farms she could get hold of in the Lake District and handing them to the nation. Otherwise it is likely the area would have been exploited and ruined by development. My own annoyance about the neglect of women's talents in science is based both on the unfairness of their treatment and secondly, on the manifest evidence of our age when women are qualifying way ahead of men in universities around the world.

Consider, for most of the twentieth century there were only eleven or twelve women who had won a Nobel Prize for Scientific Research. And don't forget that *three* of those prizes went to *one family*, the Curies. Marie got two and her daughter Irene got one. Thin pickings for the rest. In the last few years the numbers have moved up, slowly but surely. One of our triumphs was the Nobel for Elizabeth Blackburn, born in Tasmania, educated in Melbourne and queen of the telomeres, those caps at either end of your chromosomes that keep your genes healthy if they are not diminished by age or stress.

One of my favourite female Nobel laureates was Dorothy Hodgkin of Oxford. Some say she was worth *three* Nobels. She was one of the pioneers of X-ray crystallography (established by Laurence Bragg of Adelaide, the youngest person ever to win a Nobel, remember) and revealed the structure of penicillin and vitamin B12. She was also a vigorous campaigner for peace through Pugwash, an organisation set up by Bertrand Russell, and when I went to lunch at her farm north of Oxford, her house was packed with refugees from the vile regime in the then nation of Czechoslovakia.

But the innovation I found most charming about Hodgkin was her use of computers in research way back. She went to the Lyons Tea Houses and asked them how their computers worked keeping track of buns and other produce. They helped, she applied. Georgina Ferry, herself of Oxford and a biographer of Hodgkin, tells part of the wider Hodgkin story.

> The X-ray machines would've been down here [in the basement of a museum at Oxford University] probably next to the wall, and there was a table in the middle where people could sit and do calculations while the X-ray photographs were taking. The really interesting thing is that up against the windows, because the windows are high, ... the bottom of the window is about 9 or 10 feet from the floor, and she needed good light to mount the crystals before she could look at them in the X-ray machine. So there was a gallery up there with a ladder leading up to it. And up there she had microscopes, polarising microscopes, which she would

use to mount the crystals. She would have to climb up the ladder with her crystal and a piece of glass fibre and put them under the microscope and get the crystal oriented the way she wanted it under the microscope and stuck with a bit of shellac, a kind of glue, to the end of her glass fibre. And then she'd have to climb back down the ladder again with this crystal on the end of a glass fibre to put it in the X-ray machine to take the X-ray photograph.

Now when we talk about crystals, you might think of a big chunk of something like you might see in a museum, but the crystals Dorothy was working with, a lot of them, were smaller than a grain of salt. They were absolutely tiny because it was very difficult to get these organic materials to crystallise. So you can imagine you've got a thing the width of a hair and a crystal smaller than a grain of salt on the end of it and you're climbing backwards down a ladder and, certainly by the 1940s, she had already suffered from arthritis so her hands were quite deformed. But they say she never lost one.

It took a while, but eventually they managed to get crystals of penicillin. They thought that once they understood the structure they would be able to synthesise penicillin. In fact, it didn't turn out like that. The easiest way to make penicillin is still to grow a lot of mould in great big tanks and to extract the penicillin from it, but having understood the structure it has made it possible for the pharmaceutical industry to make new kinds of antibiotics based on penicillin but not exactly penicillin. So knowing the structure was extremely important because the war was on and the treatment of soldiers in battle was going to be a big area for this. More soldiers died of infections in the battlefield than died in the front line, so there was a big Anglo–American effort going on. But Dorothy was really the one, she was of that effort, she was the one who was working on the three-dimensional structure, with some colleagues elsewhere, but most of the work was going on right here in this room and indeed she solved the structure at the beginning of 1945.

I think the final solution was published in 1956 and this really made the world sit up and take notice. Penicillin had been enough of a triumph but vitamin B12 ... Lawrence Bragg said it was like breaking the sound barrier, it really was a tremendous achievement to have solved that structure. It was the biggest molecule to have been solved at that time.

RW: Dorothy Crowfoot Hodgkin — if you were to see her walking slowly down the street in Oxford, you would think it was just another old lady, slightly scruffy, with hands twisted by arthritis and the saddest eyes. Why that wistful look, one wonders? But Dorothy Crowfoot Hodgkin is not the least bit ordinary. Apart from the Order of Merit, she has attained every scientific pinnacle you can think of: Fellowship of the Royal Society of London and of the Australian Academy, winner of the Royal Society's Copley Medal and the Nobel Prize for Chemistry in 1964, which she didn't share — she won it outright, on her own. I could go on. She was Chancellor of the University of Bristol and President of Pugwash.

Hodgkin, who with Bragg (father and son) and John Desmond Bernal helped invent two new fields of science: the investigation of crystals and molecular biology. There have been twenty-five Nobels in the field so far, not least Watson and Crick. We know Watson and Crick. But the names of women I have managed to broadcast in *The Science Show* are not known, despite having done astonishing work. Some know Hodgkin was Margaret Thatcher's tutor at Oxford and gave young Maggie assignments she couldn't cope with — thus turning her away from chemistry to law and then politics!

Another female Nobelist with an astonishing career was Rita Levi-Montalcini. Her gong came when she was 100 and still serving in the Italian Senate. Here is Lyn Beazley, herself a nerve biologist and, at the time of recording, the Chief

Scientist of Western Australia, giving a summary of yet another formidable mind.

> Her motto is: You shouldn't fear difficult moments. The best comes from them.
>
> One of four children born into a loving and traditional Jewish family in Turin, the first of many 'difficult moments' was her father's objection to women pursuing careers. But she won him over and graduated in medicine from the University of Turin, specialising in neurology and psychiatry.
>
> The next difficult moment ended a brief appointment in Belgium when the Germans invaded in 1940 and she fled back to Italy, but not to a welcome reception, as fascist laws prevented Jews from practising medicine or working in universities.
>
> With the end of the war, Levi-Montalcini joined the staff of the University [of Turin]. Then, out of the blue came an invitation to join Viktor Hamburger at Washington University in St Louis. She planned to stay for only ten months but stayed thirty years.
>
> Hamburger's observation in chick embryos had triggered her work into the death of immature nerve cells if they are deprived of their target, the developing limb. When Hamburger and Levi-Montalcini teamed up their first move was to follow up their earlier studies and replace the limb bud by another tissue and see what happened. They tried a seemingly unlikely tissue, from mouse sarcoma, a tumour of connective tissue, following up clues from other work in their faculty. They came up with the same results when the sarcoma was placed at a distance from the limb, showing that the signal was a diffusible substance that could reach the nerve cell bodies via the bloodstream.
>
> At this point the story moves briefly to South America and from living animals to tissue culture. Levi-Montalcini took two sarcoma-affected mice, nestled in her handbag, to

Rio de Janeiro to conduct a crucial study using techniques developed there.

The outcome was sensational.

Levi-Montalcini watched as the clump of immature nerve cells grew a halo of nerve fibres that was most dense towards the sarcoma cells. It was even clearer evidence for a diffusible nerve growth factor. Levi-Montalcini: 'The discovery was the highlight of my life. I immediately understood its importance. I was opening up a whole new scenario.'

So what was that scenario? It was that, against the accepted dogma, the development of the brain, with all its exquisite complexity, is not rigidly controlled and is not exclusively predetermined by genes to be read out in an autonomous way. Rather brain development is an interactive sequence of events with feedback to keep some nerve cells alive and steer their nerve fibres to their correct locations. At the same time, other nerve cells do not receive the right signals and die. The hunt was on for the structure of the nerve growth factor.

Levi-Montalcini sought the help of Stanley Cohen, a biochemist a few rooms away. They needed first to find a tissue that was a good source of the mysterious factor. They found it in moderate amounts in snake venom and, after a systematic search, hit the jackpot in the salivary gland of the male mouse.

The protein nerve growth factor was sequenced and Stanley Cohen went on to discover epidermal growth factor. He shared the Nobel Prize with Levi-Montalcini in 1986.

So what is Levi-Montalcini's secret of long life and her tremendous vitality? She told the press on her one hundredth birthday: 'No food, no husband and no regrets.'

The list could go on and on. Maths does not have a Nobel, but it has a Fields Medal, considered the equivalent. No woman had won this until last year. Maryam Mirzakhani, Iranian born and now at Stanford, won and made history. I hope history gets repeated.

Then, hallelujah! — Oxford has appointed its first-ever female Vice Chancellor, Professor Louise Richardson from the University of St Andrews, an expert on terrorism. Knowing Oxford as I do, she will be ideally qualified to deal with recalcitrant dons. And the next bastion: *The Economist* also has its first female editor, Zanny Minton Beddoes. What a name, what a brilliant brain she is — she's appeared with Geraldine Doogue on *Saturday Extra* more than once. At last real talent is being recognised. What happened to it before?

So much for the stars. Other female stars, on the way up, are everywhere. At my own university (University of New South Wales, UNSW, where I am a visiting professor), there are three young women breaking all records. Veena Sahajwalla is a professor of engineering and made news by taking the carbon from old tyres and using it to make steel. She has thus recycled well over one million tyres. Now she is finding new life for defunct electronics. *The Australian* newspaper nominated her work as an innovation hub.

Emma Johnston is a marine biologist. Her discoveries from Sydney Harbour to seas around this country led to her being awarded the first ever Nancy Millis Medal by the Australian Academy of Science. She combines brilliance and inspiration as a speaker with solid research of huge significance. And now she has a Eureka Prize for the promotion of science.

Michelle Simmons is one of the youngest to be elected a Fellow of the Academy of Science and recently she became a Fellow of the Royal Society. She has children (and is married to Thomas Barlow whose book on innovation myths was mentioned previously). She runs the quantum computer outfit at UNSW and this year the work of her quantum department generally has been taken up by *Nature* to publish its own journal under *Nature*'s auspices.

Away from Australia and in another age group, I keep coming across schoolgirls well under way as researchers. Here are two examples from the AAAS meeting in San Jose in 2015. You can't see their lively faces and exuberant confidence but, take it from me, both are talent on stilts.

Alexis Forster: I am one of the top five winners of the Nebraska Junior Academy of Science, and that is why I am here.
RW: It says [in the program] that you bury your pigs. Tell me.
AF: I had three [dead] pigs and I kept one on the surface, I buried one under 5 inches of soil, and one pig under 10 inches of soil. And every day I'd unbury my pigs and would pull ten samples of hair and I would view the hairs underneath the microscope. What I was looking for was the root band, which then I could measure and determine the time of death.
RW: The root band, meaning the part of the hair?
AF: Yes, at the band of the hair you will see a dark mark that will grow and that is the band. No one knows what causes it yet, but hopefully scientists can figure that out.
RW: Right, so then you could look at the hair itself and work out when death occurred?
AF: Yes. And the different depths actually preserve the bodies differently, so decomposition rates will be increased if it is exposed to the weather or if it's protected from it. So, the pig on the surface decomposed a lot quicker than the one under 10 inches of soil. The one on the top surface had a longer root band.
RW: So this sort of investigation could be useful for forensic science?
AF: Yes, it can.

Iman Mahoui: I'm a senior at Eman Schools at Indiana.
RW: Right, and you are attached to the Indiana Academy of Science. You're still at high school, Iman, are you?

IM: Yes, I am a senior, and I actually did this research when I was a sophomore and a junior as well. So I've been in the research buzz for a long time.

RW: Just to explain, this is targeting brain tumour stem cells, so in other words we are talking about cancers, not benign stuff. I didn't know you could actually target them. How do you do it?

IM: Basically you have brain tumours that have a bunch of different cells, what is called a heterogeneous population of cells. And what you can do is you can isolate specifically the stem cells, so they are a small subpopulation, and then you can, through different natural antioxidants, specifically target only the stem cells.

RW: And then when you've targeted them, what do you do?

IM: Kill them.

RW: Kill them, wipe them out?

IM: Yes.

RW: How easy is this to do?

IM: It's actually very difficult because the research that I'm doing right now is in vitro, and one of the reasons why they are so pesky is they are actually located deep inside the brain tumour. So they found this little area that is low oxygen and high oxidative stress, and then they thrive in that region, so that makes them resistant to radiation, all forms of chemotherapy, and that's why they are really hard. And it's actually really interesting because some patients go into remission, they think their tumour has gone after they've undergone countless rounds of chemotherapy, radiation and even surgery, but what happens is people don't notice the small, minute percentage of cells that are left that are brain tumour stem cells and then they go ahead and they just decide we are going to propagate a new brain tumour that's exactly the same as the other one but it's more resistant to the chemotherapy.

I think that people can start to think a little bit out of the box and come up with more creative ways besides just picking interchangeable chemicals that they can just throw at these brain tumour stem cells and maybe look to nature, natural antioxidants, that are good for you, that are healthy for you and that can be applied realistically. Because the blood–brain barrier, it is really, really tough, and this is why neuroscience, especially neuroscience research, is hard because you really have to come up with ways that you can find substances that can break that blood–brain barrier. And these natural antioxidants can, and then they can improve the quality of life for the patient, because everyone knows the horrible side effects of chemotherapy.

What I found in my research was that you could actually decrease the levels of chemotherapy and increase the natural antioxidants that you are feeding intravenously and then you can improve the quality of life of the patient because they don't have to suffer through those side effects anymore.
RW: Iman, how on earth are you doing this work while still at high school?
IM: Coffee, lots of coffee.
RW: Just give Muslim schoolgirls half a chance and watch them take off.

Convinced? Female talent needs to be encouraged because one, it is there and doubles our capacity for innovation and discovery; two, women love the research and need to be enabled do what men have enjoyed for centuries; three, there may be a different approach by women to both the institutions and the R&D (I am not sure what it is but it could be worth finding out); and four, how dare anyone block a woman or girl brimming with inspiration on the basis that there are no female loos or the regulations won't permit? Make the toilets shared and change the regulations.

Two years ago, at the Academy of Science, Dr Brian Schmidt, our cosmological Nobel laureate, was giving a lecture summarising our achievements in astronomy over the past sixty years (the then age of the Academy). He came to the story of Ruby Payne Scott. She was with the CSIRO and may have been the first woman exploring the use of radio observation and she also did famous work on solar flares. After six years, someone suspected she could be married and read the riot act about married women being ineligible for senior employment (typing was OK). Her scientific career was over. At this point, Brian Schmidt actually burst into tears. His upset at such a travesty in 1950 still hurt. The Academy responded with prolonged applause. Times have changed — in attitude anyway — even among the crusty boffins.

Plenty of women have been denied high honours over the years — I mention Jocelyn Bell previously in this book, but there was also the brilliant nuclear physicist Lise Meitner and many other women whose gender, 'race' or modesty allowed others to take the honours.

But there is also another problem which many of us really need to do something about: CONFIDENCE. Professor Sheila Widnall, then President of the AAAS, made this the subject of her address way back in 1988. She said a point comes in a young woman's career when she stops sailing through almost without effort and starts to think 'Should I be here? I'm only me! Am I not out of my depth?' And they drop out.

Years later I was chairing a public meeting of the Academy of Science in Canberra. Our speakers included Professor Suzanne Cory FAA FRS, Professor Carol Robinson FRS, Oxford, and Professor Margaret Sheil, now provost at the University of Melbourne, formerly CEO of the Australian Research Council. All three talked about their continuing lack of confidence.

I was staggered. Two Fellows of the Royal Society, awards to fill a town hall, research of huge significance (Cory for helping to identify oncogenes, implicated in causing cancer) and yet still worried about being elevated above their station. I told them afterwards 'Keep this to yourself; just talk about how to *overcome* any wobbles, how to stay on track [there were lots of local schoolgirls in the audience who looked rather grim] but don't keep emphasising feelings of failure.'

We all sometimes feel we are about to be found out; that we are not quite good enough. It is a healthy check of progress. Those who sail on regardless can be the sociopaths running half the rapacious companies or banks wrecking our prospects. Women are not often in that category. In science their progress since *The Science Show* began has been one of the shining achievements in the field.

On many an occasion I am a judge of speakers, researchers and the like. When we have a short list drawn up I often note, 'Look, they are all women! Aren't there any boys we can put in just to give variety?' We lob in a young lad. He rarely wins. Women are on a roll. May it continue.

4 Here Come The Animals
In love with Betty the Crow

Richard Dawkins and I have several things in common: Balliol College, annoyance with religion, loathing of noise, and our reading when little. We were beguiled by Hugh Lofting's Doctor Doolittle character who was able to speak to animals. Critics of Richard may be surprised to hear of his somewhat sentimental predilection; I, of course, am as sentimental as a Ginger Rogers and Fred Astaire movie (loved them) and my choice would surprise no one. The odd thing is that we did not, when older, take the obvious, romantic course and see animals as quasi-humans. No, we both prefer creatures to be themselves and we relish the subtleties of their behaviour. Anthropomorphism be damned. But a chat with a border collie or a macaw would doubtless be instructive and I have demonstrated this in my only published novel: *2007: A True Story, Waiting To Happen*. It's allowed in fiction.

On the other hand, we do interact with animals. I feed birds on the weekend and the carnivores, such as magpies and crows, are with me in a flash, singing and demanding and looking me in the eye. But once the meat is offered, away they go on their own project — I'm dismissed. We watch them quietly from a distance, both the more up-front corvids and the less direct parrots and other seed-eaters. They are our neighbours, not really our friends. And that is fine.

The theories attached to animals have adjusted with the times. Descartes, that horrid torturer, not only did cruel things to cats, he was also convinced animals are mere machines, robots with predictable and limited reactions. Thought is not there, he insisted.

Another of my favourite reads when young was Konrad Lorenz. His books, *King Solomon's Ring* and *Man Meets Dog*, read when I was nine, had a profound influence on me, though I did not realise it at the time. I was reminded of this just before I started *The Science Show* when, in 1973 Lorenz, Niko Tinbergen (who taught Richard Dawkins at Oxford) and Karl von Frisch won the Nobel Prize for Physiology and Medicine for their work on ethology (animal behaviour).

The field was moving beyond cute descriptions of beasts, as David Attenborough told me, merely doing the three Fs: feeding, fighting ... and the other one. Now we were seeing real science being done beyond the Victorian tradition of natural history. From the mid-eighties we were watching animals *learn, solve problems, and* THINK.

* * *

Some of our *Science Show* series discussed this, especially those of Tony Barnett, professor of zoology at the Australian National University (ANU). He wanted to warn us about a) assuming too much from what you thought you were seeing in animals; and b) drawing a line too readily to human behaviours. We are not like cuckoos, chimps or meerkats, though we have parallel evolutionary histories way back.

> If you want to liken human beings to animals I can offer you the female hornbill, a very odd bird. During the breeding season she settles in a tree trunk and while she is on her nest, with her eggs, she's walled up. The male feeds her through a hole. The wall looks like the barrier surrounding a harem.

The cock bird's conduct seems to be an extreme of male chauvinism, at least if you insist on describing remote species as if they were human. But there's a snag. Careful observation reveals a different picture. It's the female that builds the wall. So perhaps we should call her a domestic tyrant?

Some writers look around the animal kingdom until they spot something that captures their fancy, but the hornbill seems not to have appeared yet in any of the modern bestiaries. Rather surprising, with skilful distortion, it could be used to support either male chauvinism or feminism.

But apes and monkeys are better. Here's a zoologist, Peter Chalmers Mitchell, writing in 1915 during the violence of the First World War. At that time some people were trying to explain war by what they said was the 'beast in man', even to justify it:

> 'Zoological analogies are presented to us, not as literature but as scientific fact and human conduct is explained after a visit to the monkey house at the zoological gardens.'

Even today, energetic people go on trying to reduce humanity to animal life and we're told that this is science. Often it is still a matter of monkeys. Baboons used to be popular but bonobos, or pygmy chimpanzees, now seem to be flavour of the month. This, from ethologist, Frans de Waal, is typical.

> The bonobos behavioural peculiarities may help us to understand the role of sex and may have serious implications for models of human society. Just imagine that we had never heard of chimpanzees or baboons and had known bonobos first, we would, at present, most likely believe that early hominids lived in female-centred societies in which sex served important social functions and in which warfare

was rare or absent. In the end perhaps the most successful reconstruction of our past will be based not on chimpanzees or even on bonobos but on a three-way comparison of chimpanzees, bonobos and humans.'

That's a quite restrained comment, though confused. I don't believe for a moment bonobos will tell us anything useful about human societies but at least de Waal doesn't say anything completely wild. And he acknowledges an important fact, a confusing fact — our two nearest relatives, chimpanzees and bonobos, differ from each other in their social lives.

Unlike chimpanzees, bonobos copulate frequently often face to face and both sexes go in for a lot of genital rubbing with members of the same sex. Among chimpanzees males are dominant but among bonobos females are dominant. They have right of way and they get first go at food. The female bonobo also protects her son from other males.

Dear Tony Barnett. He was so delightfully querulous. He did acknowledge evolutionary connections with all these animals but said, correctly, that we are so much more flexible, behaviourally, and left our ape forebears behind some *millions* of years ago. An even bigger question is why we are so fond of and integrated with animals? More than half our households contain pets; we spend billions on their food and accoutrements. Why? And why are we equipped mentally to bond with so many different species? The connection is thousands of years old as Jonica Newby noted in *Animal Attraction*, showing that distant, isolated communities still welcome a range of 'friendly' species.

You may answer 'child substitutes', 'love of nature' or some other vague construction — but, again, why is it there? E.O. Wilson of Harvard writes about 'biofilia', a state of mind that connects us usefully to our landscape, including, presumably, many of the local animals. We shall be more adapted if we know and like our surroundings. Fair enough, but animals too may have biofilia, but few that I can think of are cosy with other species. Some young mammals may link with others — dogs with cats, goats with pigs, but only if put together in the brief span when youngsters are open to socialise and mix. Adults rarely comply. They are psychologically cut off from all but their own.

Again, why are we so different? Andrew Harper, the Australian desert trekker, loves his camels; Jane Goodall loves her chimps; Tim Clutton-Brock loves his meerkats; most of us love most creatures. I am fond of snakes, llamas, cats, dogs, birds, cheetahs and ... Nemo!

Taking an evolutionary line, it is worth asking whether fondness for animals has helped us survive — beyond the brutal attitude that they are all just grub: kill 'em and cook 'em and we'll be right.

One of the most intriguing examples of increased survival from human–animal interaction came on *The Science Show* nearly twenty years ago when Dr Jonica Newby gave us the ABC TV series *The Animal Attraction*. Jonica is my partner, was trained as a vet at Murdoch University and has advised, in Melbourne, on how best to deal with pets — the best dogs for children, how to train difficult creatures, what food to buy (less tinned, more 'fresh').

Her series started with dogs and how long they have been part of the human community. At the time Professor Bob Wayne of the University of California (UCLA) had come out with the astonishing claim that genetics showed the dog has been with us for more than 100,000 years. The archaeologists howled in derision saying that no evidence of anatomically different, dog-like skeletons had been found older than about 17,000 years. Wayne replied that we did not invent fences until about 10,000 years ago so dogs could easily slip away from our fires, mate with wolves and so set the model back, anatomically, to the wild form: looking like wolves.

The row goes on. In April 2015, the journal *Science* ran a cover story about the origin of dogs and the factions jostling dates between 20,000 years and 50,000. The evidence of brain form, deducible from the inside of the skulls of dogs and wolves, plus other clues, now indicate a date of about 40,000-plus years is about right. This is enormously significant.

About 40,000 years ago something happened to humans that represented a huge step forward. We painted, expressed ideas and symbolism, had rituals, and became real communities. Many scientists explain this change through the invention of language. Something happened to help us think of words and develop the anatomy to say them. What triggered this change is another matter.

Enter *The Animal Attraction*, which featured scientists talking of evidence that wolves had followed our small travelling bands as we hunted, killed, cooked and left debris, which the interlopers then scoffed. Run this scene for hundreds of years and you can see a rather closer relationship becoming inevitable. The young wolves, during their few weeks of willingness to socialise, could become part of the human group and grow up with us. Their usefulness as early warning systems (having more acute senses), baby minders (they were happy coprophiliacs, cleaning babies bottoms with relish) and maybe, later, cooperating with our hunting in some way. Dogs adopted us; we did not really domesticate them. Spanning 40,000 years this relationship will have grown subtly and significantly. We survived with a little help from our friends. The following passages are taken from *The Animal Attraction*.

> **Ray Coppinger, Hampshire College, animal behaviourist:** I think the original wolf, or jackal-like wolves, were moving into villages or they were actually moving into the dumps around the villages. I imagine those villages were filthy places. I don't imagine they had many sanitation codes back in those days. But anyway, there was an opportunity for

wild animals to gain entrance to those villages; they might have been after human waste products, the latrines. And it's interesting there was this food source there and the only behaviour that you needed was to be tame enough, to be bold enough, to not be so shy of people or other animals.

Those animals that did had a selective advantage over their brothers, if you will, who were too shy to do it. And so I think that the evolution of domesticated animals happened kind of spontaneously and it was really done by them moving in on these villages rather than any kind of conscious idea that we are going to go out there and we're going to domesticate the animals or that we're going to go out and get a wolf puppy and bring it home and train it to be a dog. That's all foolishness.

Jonica Newby: When it all started we were dealing not with dogs but with wolves, ferocious wild creatures. So even if wolves were unwelcome invaders, why did our ancestors tolerate them?

RC: The question of whether wolves are nasty creatures or not is another one of those assumptions that have bred in mythology. Wolves really aren't very nasty creatures. In fact the best place to see and photograph wolves around the world is in human dumps. Set up a dump and they will be right there.

First of all I think that most of the early animals moving in there probably weren't even detected working the dumps. One morning they wake up and they get themselves a very tame batch of dogs, or call them wolves if you want, that are living in and around those villages.

* * *

Juliet Clutton-Brock, Cambridge, writer on human-animal relationships: I think it actually had a dramatic influence on the hunting because people in the Old Stone Age, the

Palaeolithic, they hunted with big hand axes, stone tools and presumably they rushed up to the animals and killed them or killed them once they'd fallen into pit traps. But during this Middle Stone Age, this period when the first dogs appear about 12,000 or 14,000 years ago, there's a dramatic change in the kind of tool that's used. Little flint points were presumably built into wood and were used as projectiles so that animals were killed by people throwing projectile points at them. I believe that this change occurred because the people at that time had already got trained dogs who could run after wounded animals and bring them down or who could track dead animals. Before you had this partnership between humans and animals it wasn't really possible to do that; you had to get right up to the creature and knock it on the head.

I think the change in the Stone Age cultures, from the Old Stone Age with the Palaeolithic hand axes to microlithic hunting techniques certainly did have a very big effect. The other big effect was that it hastened the overkill of the large mammals — that is mammoths, bison, and deer. And of course having dogs that would help in this hunting process would have quite a dramatic effect on the slow dwindling of these species, which later seems to have led to their extinction, although there is a considerable controversy about whether it was entirely climate change or human overhunting. I personally certainly believe it was human overhunting. And I believe that these early dogs had a very big impact on that.

That word 'tame' is instructive. I alluded to a period when mammals are more open to being social, before they quickly mature and a hostile wall comes down making them aloof to all but their own kind. A sort if tameness is brought about by neoteny, a mutation that *extends* this playful friendly period, sometimes to old age. Dogs are puppy-like as a result of neoteny and even humans may have evolved in the same way. Which is why we are 'naked apes' and have finer bone structure and less brutish faces. And why girls and boys 'just wanna have fun', even when well grown up. But what about that merging of mental process? Are we more doggish? Are they human-like in their responses?

In recent times, as University of British Columbia psychology professor Stanley Coren told me, we were able to breed dogs that were sensitive to our expressions and habits and that seem uncannily in touch with our minds. Yet all this can be explained in other ways. Yes, there is empathy, but they are still quite separate life forms, deep in their dogginess. An example of the confusion is the concept of guilt in your dog. It ain't so!

> **Stanley Coren:** If our hypothesis is correct then dogs have all the basic emotions — fear, anger, pleasure — but they don't have all those social kinds of emotions, so dogs don't feel guilt.
>
> Of course I know a whole bunch of people who are going to say, 'No, no, we had this really new white carpet and when we came home Lassie had redecorated it in earth and sunshine colours. And when we walked in the door she started slinking around because she was guilty. She knew that she had done something wrong.' Well, that's not true. What's really happening is that Lassie has learned that when you are visible and that stuff on the floor is visible, then bad things happen to puppies. And so she is not feeling guilt, she is feeling fear; she's afraid you're going to drop a piano on her head.
>
> I was challenged on this at one point. I had a national TV show in Canada on dog behaviour; it ran for about ten years. Somebody challenged me and so I gave a demonstration in which I got the owner to go out of the house, and while the dog watched I went into the kitchen. I threw the trash out on the

floor. Then we simply brought the owner back, and she went slinking off because it was not guilt — she knew, you know, 'That stuff is on the floor, I can see him, I'm going to die!'

They do read our emotions. In fact there's wonderful research that has come out of Emory University, which shows that in fact a dog's brain lights up in different places in functional MRIs when they hear a positive note in a person's voice and when they hear a negative note. And those are exactly the same places that light up in the human being.

Then came the birds. The first sensation was in Oxford when Betty the Crow did her bit. Long after Jane Goodall changed ethology by showing how chimps used tools, the Oxford group were seeing how much crows could do the same. They set up a cage in which male New Caledonian crows had access to tubes of food plus wire hooks, the only way the meal could be extracted from the tubes. The male crows duly used the hooks and got fed. But someone left the video running overnight. Was it a student?

The following day, playing it back, they saw Betty, with no hooks but some food left in the tubes. What happened next, caught on the recording, astonished the scientists. Betty found a piece of straight wire. She bent it to exactly the right angle to do the job, put it down a tube and gained her dinner. *This was not supposed to happen!*

The paper on Betty was published in the journal *Nature* and there she was on the front pages of newspapers around the world. You can still Google her doing her stuff making tools.

Professor Alex Kacelnik: We gave [the crows] a piece of food in the bottom of a tube. The food could only be retrieved by a hooked wire but we offered the animals a straight wire and a hooked one and they had to choose one or the other; and we are simply saying that more than random they chose the right one.

So what happened was that the male stole the hooked wire and took it away, and then Betty was faced with a straight wire and could not get the food. And then she solved the problem in a novel way.

RW: She actually bent it?

AK: Yes, she actually in a sense did more than that. Because you realise you have two hands; if you were to do it you would grab one extreme with each hand and bend it. What she did was, wedging one tip in a crack in a tray and picking at the other extreme, pulled until it had the right angle. And you can see the many different tools that she's made and they all have approximately the same angle; she knows what she wants to get.

What Betty's doing is obviously solving a new task by generalising from things that she's done before. This is to me one of the most exciting things ahead of us — whether these animals are generally clever for many different tasks or they have a dedicated brain; they have a 'kind of mind', so to speak, which is very good at a certain category of problems but is actually perhaps completely ordinary with respect to everything else.

This research has been taken further by Nicky Clayton's team in Cambridge. One of their experiments (mimicking Greek legend) was to put tubes of worms floating in water out of range of the birds' beaks. There was also a pile of small pebbles provided. The crows looked at the food, then the pebbles and in no time used the stones to raise the level of the water to yield the lunch. They were applying Archimedes, and their Eureka moment was getting the food. Professor Nicky Clayton in Cambridge:

> It turns out that the bird brain was massively misnamed and has been renamed. We now know that there are whole areas of the bird brain that are cortical-like. They don't have the six-layered cortex structure that mammals have because

birds have a different type of brain. Theirs is what's called nuclear, which means it has little conglomerations of cells. The analogy would be that the mammalian brain, being layered, is a bit like a bacon, lettuce and tomato sandwich with all its layers, whereas the bird brain, being these sort of little clusters of cells, would be more like a pepperoni pizza. They have a very different type of organisation, but nonetheless birds have a cortex-like area.

Now, the really fascinating bit about that is it turns out that members of the corvid family, including Betty the New Caledonian crow and our scrub jays have very large brains. Their brains relative to body size are as large as those of the great apes. And what's more, the area of their brain enlarged is thought to be the equivalent of the neo-cortex, the frontal areas of a mammal's [brain], and that this increase in the crows mirrors the increase or expansion you see in the frontal cortex of apes, including humans.

That's one reason why we think that actually the crows at least have the brain machinery for this intellectual behaviour.

RW: Could it be that other birds — parrots, which also live a long time — could turn out to be these super-brainy creatures that we've underestimated all along?

NC: Absolutely.

And now Kacelnik's group has turned to Australian sulphur-crested cockatoos, concealing their food behind five separate locks — which the birds quickly examine and open to get fed. However the scientists mix up the differing locks, the birds still tackle them without hesitation. It is like a game. The reward is food.

Here are creatures with bird brains analysing a situation, seeing the challenge, and solving a problem using insight and planning. This was unthinkable way back when I began *The Science Show*. We have learned that the more you find out about animals, the more subtle and interesting they turn out to be. And the consequence of this research is really profound.

Our natural world is far more complex and fascinating than we ever imagined and the level of this complexity is only now being revealed. The world around you, even the dog by your feet, should never be underestimated.

Isn't this obvious? No. The world is being trashed at a rate that looks suicidal. You are more fascinated, some of you, by screens with apps and the habits of confirming your own attitudes using them. Surveys in America (especially) and Australia show an astonoshing ignorance of evolution and biodiversity. If this is not changed we risk losing living things whose attributes are staggering — yet we'll never know because they haven't yet been discovered or studied. Without knowing the natural world around us we shall become the first generation alienated from where we originated.

We can talk to some animals, though not in the way Doctor Doolittle was able to. But we can now understand how many of them think. There are techniques of enquiry and interpretation revealing worlds that even the most unromantic of investigators find to be remarkable. Film-makers such as David Attenborough, Brian Cox and Alice Roberts have shown us some of these 'wonders'. We used to do the same in Australia through the superb work of Dione Gilmour, David Parer and Liz Parer-Cook in the ABC Natural History Unit. Now that is over, their cameras are shut down. This is both sad and unforgivable.

5 There Goes the Clever Country
What happened to our bright ideas?

When Barry Jones became minister for science under Bob Hawke, he was keen to present a few promising innovations to us, to show how a new, brain-based approach could take over from our traditional dependence on commodities. He chose the natural sunscreen to be extracted from corals. It seemed a good idea at the time.

Corals are little animals. They have algae inside them to provide food (carbohydrates) through photosynthesis. This means the algae must have light. But tropical waters, combined with exposure at low tide, offer up the tiny seaweeds to harsh glare, so the clever polyps are able to secrete a film of protective screen in much the same way that we put SPF 30+ on our pink skins.

Now, thirty years later, where is this 'natural' sunscreen? And where is Ralph Sarich's famous rotary engine we were hearing about at roughly the same time? (He now invests in property — same ol' story!) Inventions often have a tricky time being realised. But Australia seems to have blockages beyond those experienced in other OECD countries. Not enough money (are we really an impoverished nation?), bad connections between academe and business (notorious), and that bovine recidivism to commodities allowing you not to have to bother. Wave power, geothermal energy, oil from algae — even solar power — have languished in a nation that holds the record for solar efficiency and wave technologies.

This affects all of us because scientifically based research is the basis of a nation's wealth. Real wealth, not mere paper fortunes. My 2011 interview with Jonathan Haskel, professor of economics at Imperial College in London, showed how much (39 to 50 per cent?) of our GDP is essentially R&D based. Our top manufacturing goods are headed, among others, by medical technology — all those apnoea machines and vaccines: brain-based products.

And there is a myth (yet another one) about the twenty-first century entrepreneur idealised by Steve Jobs, standing there in the spotlight in his slim black gear, holding the latest iThing. Professor Mariana Mazzucato from the University of Sussex points out bluntly that *all seven* of the iPhone's key technologies come from state-funded campus research, from GPS to touch screens (the latter, in part, care of the CIA!). Her book, *The Entrepreneurial State* — debunking private versus public sector myths — is full of such startling reminders that our scientists and engineers, from universities to CSIRO, have had their achievements neglected. No wonder many give up.

We do have our successes. This is how Professor Mark Dodgson sums it up.

> We can be wonderfully innovative in Australia.
> When we get up in the morning we might use our dual-flush toilet, have a breakfast of Vegemite and an Aspro to overcome the hangover we got from drinking from the wine cask the previous night, clean our shoes with Kiwi polish, go for a swim in our Speedos, find our way using Google Maps on Wi-Fi, make our tea using Zip instant hot water, do some woodworking on our Triton workbench and some gardening with Dynamic Lifter and a two-stroke lawn mower bought with polymer bank notes. In so doing, we are immersed in Australian product innovations. When we add

some of the organisational innovations in the public sector, such as the Flying Doctor Service or surf clubs or the Higher Education Contribution Scheme, or service innovations such as Macquarie Bank's funding of infrastructure, then we have much of which we can be proud.

And there is Wi-Fi as Mark indicated, the technological principles that makes it possible were found almost by accident by CSIRO researchers looking for mini black holes; Gardasil vaccine; lithium for depression; Martin Green's 40 per cent efficient solar cells and many more. But, realising these bright ideas has been a struggle. How much more could have been achieved if only we had the entrepreneurial spirit of Israel or even America?

Here, from a series summing up our innovation history and present predicament is Mark Dodgson, professor of business studies at the Business School, University of Queensland. He is, in my opinion, the top man on the topic both in Australia and in other parts of the world. He wrote a *Science Show* series for us and it is worth quoting at length because of its importance today and the power of his argument.

> Innovation is hugely important, and while we have much to celebrate, we can and need to do more to improve our innovation performance. It is the successful application of new ideas — my definition of innovation — that leads to the productivity and economic and social welfare that we seek and need.
>
> Innovation matters.
>
> We can do better and we need to do better. All the world's economies are becoming more knowledge-intensive; that is, the value they create is produced by knowledge rather than investments in factories and equipment, and innovation is what derives economic and social value from that knowledge. We need to more consistently produce returns from our ideas

and to ensure those returns are environmentally sustainable. We lack the incentives and skills to create the normality and regularity of the innovations that are the motor force of productivity in the knowledge-based economies of the future. Just over a decade ago Australian productivity went from growing substantially faster to growing substantially slower than the OECD average.

This will be our downfall unless we address these shortcomings, especially as they apply to the immense challenges of developing sustainable industries. Innovation to deal with environmental crises may well be too late as we may have passed the point of no return. We have the diversity, creativity and energy that are conducive to innovation, but there is much to be done, especially in improving the ways we manage innovation and our collaboration between firms and between business, research, policy-makers and community groups.

Any history of Australian innovation would be incomplete without reference to Arthur Bishop. He was born in 1917 and died in 2006, and displayed many of the quintessential characteristics of the inventive entrepreneur. He possessed immense intellectual curiosity and passion about his discipline, and the craft of mechanical engineering, and displayed dogged persistence in the face of overwhelming odds. In his career he confronted the indifference and arrogance of the British military in the 1940s and the power, introspection and occasional dishonesty of the US car industry in the 1950s and 1960s. Bishop's journey is testament to the great returns and personal costs of a career in innovation.

Like many inventors, his primary motivation was understanding and improving things; in making things work.

His engineering career began in earnest during the war with the building of the Beaufort Bomber, used extensively by the RAAF in the Pacific, and the only bomber ever built in Australia. The planes were desperately needed for our war effort, but supply of the British-designed bombers inevitably dried up with Britain's other pressing priorities. When parts such as undercarriages and propellers could no longer be supplied, the challenge was on of building aircraft ourselves in a country that did not yet make cars.

In a highly impressive logistical feat, an aircraft factory was built in Sydney that was up and running within a year. Eventually the Australian-built bomber, 700 in all with their 39,000 parts, was manufactured by an organisation of 8500 people, which was Australia's first scientifically designed company.

Bishop played a central role in the efforts to build the bomber. He designed an important innovation in the landing gear. He developed instruments better than those found in the US and Britain, explaining, 'We had the advantage of having no inhibitions, no sense of what could or couldn't be done.'

The greatest challenge of his career, however, lay in his attempts after the war to introduce his radical new design of variable-ratio steering gear to the US car industry. It became a ten-year struggle as he engaged in vicious political battles involving staunch supporters and implacable opponents.

His difficulties in America were affected by that nation's almost complete ignorance about Australia. He was an outsider, an outspoken and opinionated foreigner, telling the world's most powerful industry he knew better than them. Bishop had to confront one of the greatest challenges of innovators; questioning existing products that were profitable and successful. His steering system was technically better, but existing systems worked, were well established and produced substantial returns to their producers, inducing highly entrenched resistance to change.

Bishop dealt with companies that had no qualms about behaving unethically, reneging on deals and infringing patents, and resorting to dirty tricks.

Eventually, his financial success derived from licences to his patents on the valves on variable-ratio gears and their manufacture, as the industry adapted to Bishop's ideas and, rather than wholly adopting them, used them to incrementally change existing products.

A restless inventor throughout his life, he developed heaters, car trailers, air conditioners, a concrete mixer and a rotating house. His name appeared on eighty-five patents, including twenty-five on the design and manufacturing of steering gears. He invested hundreds of thousands of dollars of his own money back into his ideas.

Bishop's career tells us a great deal about the trials and tribulations of the innovator and the rewards that ensue when good ideas are persevered with. Many of these apply to innovators in all countries, but Bishop also highlights very clearly the difficulties of Australia innovating on a world stage. Our domestic market is so small that innovations of great significance need to additionally confront the huge costs and challenges of internationalisation.

Now let's turn to an innovative policy-maker. Just as the Hawke–Keating governments in the 1980s transformed our economy by liberalising, floating the dollar, reducing tariffs and negotiating the accord that tied wage constraint with social advances, our innovation system changed as well. And it needed to.

One person who deserves much of the credit for changes in our innovation system is Senator John Button, who tells us of his time as minister of industry in the 1980s in his entertaining autobiography, *As It Happened*. He holds no punches on his views about Australian industry at the time. Here are his views on the factories he'd visited: 'Some were industrial museums, others industrial-relations bear-pits. I hadn't seen many to which you'd want to take a discerning overseas visitor.'

He saw industry associations demanding subsidies and tariffs, and unions locked into protection. He saw short-termism. He points out that in the late 1980s Australia processed less than 5 per cent of its wool. He argued 'the prevailing view in government, and most of industry, was that if we needed new technologies they could be licensed from overseas'.

To respond to a crisis in the automotive industry, in 1983, Button set up a council of all the stakeholders in the business: manufacturers, component suppliers, importers, dealers, unions, consumer organisations and public servants. He was astonished that they had never, ever met before as an industry.

Button wanted to support what he called 'sunrise industries'. He wrote that, 'By 1987 I'd tried to explain my thinking by coining the phase "a productive culture". I meant a society in which there was collaboration between various economic agents to support wealth-generating businesses, particularly new ones that employed skilled people: support, for example, from universities, banks and other financial institutions, export agencies and governments. It was about linkages and networks, the sort of thing I'd seen in Sweden, Germany and Japan. In Australia the idea seemed a vainglorious ambition.'

I think now that he was successful in many, but not all, of his ambitions.

Any discussion of innovation in Australia has to include the CSIRO, the Commonwealth Scientific and Industrial Research Organisation.

The fifties and sixties were good for the CSIRO. There was a climate of optimism about the potential for science to deliver economic growth. When myxomatosis was used to control rabbits in 1950 it was the first ever successful biological control of a pest mammal. CSIRO pioneered other significant breakthroughs in areas as diverse as radio astronomy, metal smelting and mechanised cheese-making. Its expertise in radio astronomy and wool science, in particular, contributed to its growing international reputation. By the 1970s, CSIRO was a large and diverse organisation that dominated the Australian scientific landscape with strong research expertise in thirty-seven different divisions.

By the end of the decade, however, we had the oil shock and the ructions caused by trade liberalisation. Universities in Australia were getting stronger in their research and began to spend more than the CSIRO. It was no longer clear to policy-makers that 'scientists left to themselves would deliver the benefits that industries and national economies needed'.

Thus began seemingly endless rounds of reviews of CSIRO's goals, governance and structure, with the result of moving it towards research driven by business needs, increased external funding and more commercial outcomes.

The CSIRO has recently had a major success in aggressively defending its Wi-Fi patents in the USA, with a windfall of over $200 million. The research that led to this result was not driven by any idea of how it would be used in the market. No one had a clue how companies might use it. By taking an elevated view of Australia's innovation system, we can see how spending is moving away from the more speculative investments, and perhaps in the case of the CSIRO, limiting its ability to come up with future Wi-Fis. We should also be concerned whether we're prepared to wait the decades it often requires for science to translate into innovation.

The reasons we aren't better innovators in Australia has been explained by a lot of factors, by a lot of analysts: poorly developed capital markets for investing in innovation, badly formulated and directed policies, an absence of large domestic markets and large high-technology firms, and a culture unexcited by entrepreneurial achievement.

My argument all along has not been that we lack inventiveness and ingenuity, but that we could become better at innovation by developing better policies, by being better managers, and by being more collaborative.

With more and more of the world's wealth created by international trade, and with more and more global trade being in knowledge and ideas, we in Australia need to have something unusual and smart to bring to the table. If you are not at the forefront you have to pay the premium prices that create trade deficits. We import four times as much intellectual property than we export. Apart from the simple economics, there is no pride in dependence. Being great at having ideas but not exploiting them is akin to being the best at producing, preparing and training sportspeople, but being unconcerned when they fail to display their abilities on the track and pitch.

An example of a lost opportunity is provided by the space industry. Despite an early start, and favourable circumstances, such as vast areas of empty land and helpful atmospheric conditions, while we have some niche interests we have no space industry. The use of satellites and the data they produce provides the basis for modern telecommunications, meteorology, environmental management, mineral-resource prospecting, defence, and even border protection. Australia had a head start in research and innovation in the field, yet we now have little control over this crucial element of modern infrastructure.

As well as learning from failure, we should learn lessons from Asia, where countries such as Korea have moved in a generation from abject poverty to considerable wealth as a result of, as they put it in Korea, moving from imitation to innovation.

Among the things we need to do is to build much greater understanding about innovation among our policy-makers, businesspeople and researchers, especially about the importance of being more collaborative.

Bureaucrats like things that are straightforward and accountable, and can't deal with the messiness of innovation, its trials and errors and uncertainties. Their viewpoints are based on rules, predictability, conformity and compliance. Their worlds do not like change and dynamism. Well, the world isn't and probably never was that certain. All elements of society are confronted by turbulent change, but the public service is the part least prepared to deal with it.

So much would improve if only the very clever people in Treasury and Finance could break out of the indoctrination they received during their economics degrees taught by people with no experience of business and stuck in an imaginary world of absurd assumptions about individual and organisational behaviour.

We will become more innovative when we become more collaborative. Whereas invention and ingenuity often result from the actions of individuals, innovation is essentially a collective activity, involving many different people and organisations. It is a collaborative endeavour. As John Button put it in his autobiography: 'The problem was not the absence of inherent abilities, but the absence of linkages.'

Times have changed, and universities are generally much more receptive to the idea of collaborating with business. Unfortunately, due to some totally misguided views in government and senior echelons within universities, these connections have attempted to be managed by technology transfer, or industrial liaison offices, which in many cases have actually presented barriers to engagement.

We are very badly served by our business schools when it comes to researching and teaching innovation. And perhaps we might expect our leading advocates of science and engineering to be more forceful in their arguments about the importance of innovation research. My jaw nearly hit the table when I saw a recent missive from one of our esteemed learned societies saying we didn't need research on innovation in Australia as such research was conducted in America.

There is a similarly bleak picture from the business perspective. Although there is much to celebrate in the steady increase in R&D expenditures in business, OECD evidence places Australia lowest among its members on capacities for collaboration between firms and between firms and higher education, and second lowest on collaboration between firms and government.

Innovation is stimulated when researchers, businesses and governments work together.

Governments can play a crucial role coordinating and facilitating the connections we need. It can play the role of connector by its support of dynamic and evolving national institutions and infrastructure, and through its educational and research programs and procurement policies it can encourage the development of organisational skills and capabilities. Around two-thirds of government expenditure on innovation goes on supporting R&D in single firms. It is time to shift the balance towards supporting collaboration.

Businesses can improve their innovation performance by appointing innovation directors with resources and influence, and by reporting to Boards of Directors on innovation performance.

My purpose has been to highlight the importance of innovation. We have a massive productivity challenge in Australia and it is really disturbing to see how the recent discussions of productivity among political and business leaders fail to emphasise how innovation accounts for most of it.

I have argued that history matters and our approach to innovation —where we develop wonderfully ingenious ideas, but often fail to commercially develop them — reflects a history of reliance and a whole series of impediments in the ways our firms, governments and research institutions are structured and behave. As with our successes in the mining industry in the nineteenth century, collaboration is essential to our future prosperity.

We have massive future opportunities that result from our diminished isolation in the digital age, where we have instant and cheap communications with the rest of the world, and the tyranny of distance is replaced by the benevolence of propinquity with the fastest growing part of the world. There are fantastically productive experiments occurring with innovation focussed on cities and local precincts within them. Emerging digital technologies and the deluge of useful data being produced and shared by infrastructure such as the National Broadband Network provide the prospect of faster, cheaper and more democratic innovation and much better public governance. But we have to get it right, and we have to get away from the views at all levels of government in Australia, and in our businesses, that innovation is an expense not an investment. This leads to the dangerous assumption that it should be cut in times of austerity, and there are worrying signs that we are cutting back post global financial crisis. Innovation, like the human organism, needs constant nourishment, and cutting off a limb means you can never grow it back.

Innovation is not the same as R&D or invention or patents or new technologies. It might involve all these things, but in essence it is the successful application of ideas across the wide range of things we do and how we do them. As such, it involves everyone. As it involves us all, the nations, regions, organisations and people that will be most successful in the future will be those that are simply better at working together at combining their particular expertise and resources. Yet our current political climate accentuates levels of dissonance and polarity that are widely and insidiously destructive.

To see continuing improvements to our lives brought about by innovation we need people and organisations capable of transcending professional, disciplinary and organisational boundaries. We need better collaboration between different levels of government, between and within the public and private sectors and between what we do here and what is done overseas. It is the ability to work together that will provide the adaptability and vibrancy our society and economy needs and provide the rewards to our children as optimistic, creative and fully engaged contributors to a happy and productive Australia.

So there is our predicament. Catherine Livingstone, President of the Australian Business Council, said in 2014 that our industrial policy was 'broken'. Since then, various gestures have been made to highlight innovation including a prize to go with the Prime Minister's Science Prizes.

One of the problems is that it is so much easier and cheaper to avoid disruptive innovation in favour of incremental. Look at the BlackBerry, once in every smart suit's pocket. It cost Mike Lazaridis millions to pioneer. Then along came Apple and others and they jumped to the top of the market, making the most of the investments Lazaridis had made.

Disruptive innovation can be unforgiving. When it works (colour TV, laptops, the pill), everyone sees it and jumps to copy. Competing with the big wide world may be tough for a small country (however those in Scandinavia seem to do all right). But our innovations can also suit our nation, not necessarily be offerings to that tough world market. The potential is huge. But time is evaporating.

Prime Minister Malcolm Turnbull appreciates this. He has described our times as a great opportunity, despite the uncertainties (see his remarks on page 247). Accordingly, he has made several key appointments to his ministries to foster innovation. He has also begun to build bridges to other nations so that collaboration, so far neglected, can be encouraged. It will take time. As Catherine Livingstone reminds us that, though our famous Bionic Ear is a resounding success as developed by the company Cochlear, it took *seventeen years* to get the enterprise going and there were many snags along the way.

Our new Chief Scientist, Dr Alan Finkel, insists that we have several world-beating innovative industries in Australia, not least banking and mining and it was fitting in 2015 that the first Prime Minister's Prize for Innovation went to Newcastle University's Graeme Jameson for his work on bubbles (truly!) saving billions of dollars in the processing of mineral stocks like coal and ores.

But we need to be ready to take up the PM's challenge urgently for two reasons. First, our scientists and engineers are brimming with bright ideas which should be realised quite outside the predictable range of activities. Secondly, there is a need to be more productive and efficient; growth may be destructive in a old-fashioned sense, but the growth of clean, smart industries is another matter. Or could be.

6 Climate Science and Terrorism
Armageddon comes to science reporting

On September 11, 2001 I was at Imperial College recording interviews. It was a clear afternoon, pleasant for London, and I took the lift to a high storey to talk to a man about a dog. I knocked on his open door and went in. No sign of him.

He appeared abruptly, tall and with curly white hair. His face was sagging and he looked past me and began to sway. Without prompting he said, 'They've bombed the Pentagon, now they are attacking the White House.' We sat. I put my hand over his shaking one trying to comfort a man as if I'd known him for forty years, not forty seconds.

He explained in faltering voice what seemed to be happening. Half of it turned out to be wrong but the reality was still awful. I told him it was Bin Laden. I don't know why I was so sure.

'Look,' I said, 'you don't have to do an interview. Just sit—'

'Oh I do!' he protested, 'Now more than ever. My little dog is made from a plastic that children with allergies can cuddle without risk. On a day when they do evil, we must try to do good.'

So we recorded our talk about his safe toy and why he thought it to be important. The discussion seemed to calm him. But when we had finished he went back to his radio or TV for more of the horror in New York and Washington. The world had changed.

Outside I emerged from behind the Albert Hall where the proms were about to finish in a few days and walked to the South Kensington tube station. The station was crowded but we were told the Piccadilly line was closed because of a bomb threat. Everyone, in silence, descended to the line that was still open. I took a train to Regent's Park. Still no one talked. It was as if London was entirely stunned.

My walk along the park took me to the delightful cottage where Michael Frayn wrote his plays and novels. I knocked on his solid front door. It opened. 'Oh, I thought you might not come now ...' said Frayn, looking as donnish as ever. 'It is a very good day to record a talk about the end of the world,' I replied, rather too sardonically.

He made tea. We discussed the bombing using planes and wondered at the ghastliness of the terrorists training to kill themselves along with thousands of others. And then we recorded a conversation about Frayn's brilliant new play, *Copenhagen*, which was about to open at the Sydney Theatre Company, directed by Michael Blakemore. It was another excellent recording. I thanked him, almost a chum (having met him and interviewed him many times) and went to visit friends. That night I saw the planes plunge into the World Trade Center at least fifty times, almost as if on a loop. I and you — we would see it hundreds of times more. September 11, 2001: truly a different world, made in minutes. From that day on there were two big things to dread: there was the ruthless terrorist — and there was climate change.

* * *

When a big science story breaks you would imagine I would be joyous. Not so. When the *Columbia* and *Challenger* space shuttles blew up I groaned twice, once in sad reflection but once also in despair on how I could possibly handle the story in *The Science Show* that week. What do you do if you are on your own (we have no reporters or researchers) while teams from *60 Minutes* or *Four Corners*, with helicopters and thirty people on five continents are tracking every aspect of the breaking story?

I give it a mention, just to cover it, and then promise listeners a follow up later. When the herd has moved on in a week or two, I find someone with something extra to say about the tragedy.

The same with climate. Having put it in the first ever *Science Show*, I found follow-up reports for other shows as the research progressed. Even then we did not treat the issue as Armageddon; it was a concern, like so many other environmental worries, but not at that stage *potentially shattering*. That came later. In 1988, with the Commission For The Future (established by Barry Jones with Phillip Adams as Chairman and me as deputy), we launched Greenhouse '88, a scheme to foster debate about climate research so this nation would not be left behind as significant developments occurred. It was thoroughly science-based. We even won a UN prize for the effort.

All was steady and reasonable until the beginnings of the new century, when opposition erupted. It was partly in response to Al Gore's famous film (it got an Oscar, no less), *An Inconvenient Truth*. Then came Tim Flannery's book *The Weather Makers* and warnings from leading green groups and the Intergovernmental Panel on Climate Change (IPCC). The opposition interpreted all this as a conspiracy to impede growth and undermine the precious free market. Some (Maurice Newman) even warned darkly of WORLD GOVERNMENT (!). An odd forecast when we can barely elect local government efficiently.

All this was distilled with added bile of snake and phlegm of boar in the polemical film, *The Great Global Warming Swindle*. From 2007 onwards all pretence at nuance was abandoned. Everything hurled by the critics was 100 per cent, nothing less. 'Hoax, Swindle, Alarmist, Doomsters ...' and the leading climate advocates had their personal lives plundered and trashed from that day to this.

My reaction? I put the critics on the radio: Ian Plimer, Don Aitken, Bob Carter, Jennifer Marohasy, Willie Soon, Sallie Baliunas, Freeman Dyson, Matt Ridley, Richard Lindzen, Nigel Calder and plenty more. This was against my editorial principle of not broadcasting members of lobby groups but only scientists who'd done published research. Of the above only Soon, Baliunas and Lindzen qualified.

My lesson came soon. They all wanted to be back on the airwaves ASAP. To say what? Why, the same as before. The same phrases came out, even ones shared between several of the speakers. It was clear that this was an exercise in PR with likely training and media counselling of the deniers by professionals. The Koch Brothers in the USA (worth about US$30 billion — *each*) were funding think tanks and the wilder extremes of political parties and, I would hazard, plenty of smart publicity coaches.

Enough! They could rampage away on the current affairs programs and news, but there was too much original science with *evidence* for me to allow propagandists free rein. Eric Abetz, the Cabinet minister, complained about me on Q&A. Columnists had a field day. It goes on.

All this culminated in an interview I did at the University of Western Australia two years ago with psychologist Stefan Lewandowski (now at the University of Bristol). He had been involved in two published studies asking, 'Who are the climate deniers and how many are there?' He combined them into one study with CSIRO. The conclusions are fascinating.

Stefan Lewandowski: I became interested in scepticism generally a couple of years ago in the context of the Iraq War, and I discovered people who were sceptical of the reasons underlying the war processed information more accurately. In 2009 when there was this eruption of so-called scepticism with regard to climate change, I thought, well, let's look at this and see if these people are really sceptics. And I then did have a look at the scientific literature and at what these so-called sceptics were saying, and I discovered that in actual fact those people weren't sceptical at all; they were rejecting the science on the basis, not of evidence, [but on] some other factor.

And so I became interested in finding out what that other factor might be. And together with a couple of other people around the world, I started doing research on that, and what we basically found is that the driving motivating factor behind the rejection of climate science is people's ideology or personal worldview, their fundamental attitudes towards how a society should be structured. That is what determines whether or not they accept the scientific evidence. And specifically what we find is that people who are endorsing an extreme version of free-market fundamentalism are likely to reject climate science. They are also rejecting the link between smoking and lung cancer, they are rejecting the link between HIV and AIDS. So there seems to be something about an extremist free-market ideology that prevents people from accepting scientific evidence.

RW: However, there is also a left-wing component, because I know of a number of Marxists or ex-Marxists who would infer that the greenies who are keen on climate science are trying to deprive the poor of the world of the benefits that we have had in developing civilisation. Have you come across that kind of left-wing component of this as well?

SL: It's an interesting question. I've been doing some research for the past year or so, chasing people on the left side of politics that are rejecting scientific findings, and it turns out that that search has been extremely difficult. I've done a number of studies and it turns out that statistically you cannot detect much of an effect from the political left.

RW: There have been some extraordinary statements — I'm thinking of a couple of bankers I've met who have sat through a learned lecture by one of the most famous scientists in the world, and came out the other end and said, 'Well, of course, it's not proven.' And I don't know how much more evidence you'd want. And the former chairman of the ABC, Maurice Newman, who had been head of the stock exchange, came out with some drivel in *The Australian* newspaper about how climate science is a religion and a hoax.

SL: I think if you're driven by ideology rather than evidence then by an act of projection you have to accuse scientists of being religious in order to justify your own denial.

RW: But let me ask you that question — if you're using such easy debating techniques, you're just insulting 97 per cent of the climate scientists who agree that there is a problem, and you're doing so even though you are the kind of person who is used to talking to Cabinet ministers and the highest-grade clever experts in a field; you get the best advice but choose to ignore it. That's amazing.

SL: It's absolutely amazing, and in fact they are not just insulting 97 out of 100 climate scientists, I think they are also insulting basically the Enlightenment and everything that we've worked on for in the last couple of hundred years, which is to go from a dogmatic and religious approach to life to an evidence-based approach. I think that the dismissal of science by people who are interacting with Cabinet ministers, as you just said, is actually a very critical issue that is facing our society and we have to understand what motivates those people.

And one of the intriguing results is that neither education nor intelligence is overcoming the influence of ideology. There are some American data on this, which show that among Republicans the greater their level of education, the more likely they are to reject climate science. So in other words, educating Republicans drives them more towards denial, whereas if you educate Democrats, and you look at the effect of education on Democrats, then you find that the more educated they are, the more they accept the scientific findings.

So you get this increase in polarisation with education between Democrats and Republicans. And the same thing is true if you ask people about their rated self-professed knowledge of climate science, then people who are politically conservative, the more they think they understand the science, the more they will reject it. Ideology is the overriding variable in this.

RW: Well, there's the problem. Whether the scientists are ultimately right, and scientists can be wrong, nonetheless it has prevented climate science and the possible dire consequences of what we are facing to be discussed in the election in the United States. That's a serious problem, is it not?

SL: Absolutely, I think it is a serious problem, which is one of the reasons why I'm working on it, because we have to understand what motivates these people and how one can deal with that. And one of the things that one can do is to underscore the consensus among climate scientists about the fundamentals of climate science. You mentioned earlier that 97 per cent of climate scientists agree on the fundamentals, and that number is roughly right — it's in the nineties, 90 per cent or more. It turns out, and one of my recent studies showed this, that if you tell people about this consensus and the strength of the consensus, and if you just show them a graph that shows ninety-seven people who agree on one thing and then there are three who don't, that consensus information does shift people's attitudes.

And what I found in one of my studies is that that shift in attitudes is particularly pronounced among people who would otherwise reject climate science based on their personal ideology. So that is one of the things that I think is a successful strategy, just to keep underscoring the consensus, the fact that the scientists agree, the fact that every single major scientific organisation in the world is endorsing the basics of climate science, and so on. I think that is a very important thing to underscore over and over.

RW: And what about the number of people who are, if you like, denying it? A paper published by one of your colleagues this week suggests the number is fewer than one would have thought.

SL: Absolutely, and that has been shown over and over again. In Australia, in this survey you just mentioned by Iain Walker, the number of people who deny that climate change is happening is around 5 per cent or 6 per cent of the population. But those 5 per cent, if you then ask them how many people they think are sharing their opinion, their response is, oh, about 50 per cent. So what we have here is a fringe opinion that is held by a very, very few Australians, but they have convinced themselves that half the population agrees with them. And this phenomenon is called a false-consensus effect, technically, and that phenomenon is usually indicative of a distortion in the media landscape.

Other research has shown that in other instances, that if people develop this sort of self-inflation, where they are inflating their self-importance, and that usually is indicative of the media not doing their job properly. And there's no question in my mind that in Australia the media have done a terrible job in representing the science. And there have been a lot of analyses recently pointing out that particular publications out of the Murdoch empire are systematically misrepresenting the science, distorting it, representing things that are simply not true. That happens over and over again and is difficult to explain by any sort of random process. There must be something else going on there. And I think one of the consequences is that this fringe opinion has taken hold in public discourse.

RW: You're a psychologist and I think there is no great difficulty in trying to understand that if you're being told that the entire globe is threatened in a way that is pretty dire, you'd rather think otherwise. So isn't there a component of wishful thinking about this as well?

SL: Absolutely, totally, I think that is absolutely true. And what is very interesting about this is that there are some data to suggest that a lot of the people who deny at first glance that human beings are responsible for climate change, they actually do know that we are responsible, and it's a very funny result. This is, again, in Iain Walker's research. What he's done is to ask people, 'If the globe is warming, are people responsible or not?' And then it turns out that about 40 per cent to 45 per cent of the people will acknowledge the fact that the globe is warming, but they will say, 'No, people have nothing to do with it, it's all natural fluctuation'. Then, a minute later if you ask them who was responsible for global warming, pick a few of the following from this list, then even the people who just said it was all natural, pick 'polluting corporations', 'large industrialised countries', etc., and assign the responsibility for warming to them.

That tells us that these people actually know who is responsible and so their 'denial' is only skin deep, and I think what's happening there is this is just a tool for people to exercise their wishful thinking, to say 'No, it's all natural fluctuation' and then they can go on driving their big trucks or whatever it is they're doing.

I think there's a larger implication of this, and that is that one of the problems we've been having is that climate change has always been communicated in a doom-and-gloom fashion. And obviously that turns people off, and it's totally understandable why it would do that. What we have to find is a different way of talking about climate change and in a way that is underscoring the opportunities that come along with it when it comes to the development of clean energy, underscoring the fact that the problem is solvable, with considerable effort and money, but it is a solvable problem. And I think we have to highlight those solutions, and we have to try and highlight the fact that there are new entrepreneurial opportunities out there in dealing with the problem.

So, there are relatively few 'deniers', despite the cacophony they make, and most are prompted by right-wing political views. Their zeal comes from trying, they say, to protect the market economy and wealth underpinned by energy from fossil fuels (they have no faith in the rapid march of renewables); and in some minds, such as Maurice Newman's, they are also keen to resist the threat of world government. The odd thing is, as we shall see, there are countless new technologies that could be the basis for huge, important industries. It seems to me that it is not a cry to smash the market (beyond the outbursts of the naive few), but a change in the products being offered: away from the nineteenth-century industries of filth, destruction and waste towards smart, clean, efficient alternatives that could soon be as big as Microsoft and Google have been in the IT revolution.

My predicament is straightforward. I am a science journalist and need to report on what has actually been discovered, not someone's febrile opinion of what may be the case. You report on what is *there*, what you find when you go out and look, not what you or others make up. If I did the latter I would be helping to invent 'facts'. I would be going in contradiction to all the leading academies of the world, tens of thousands of top scientists, most governments, including Australia's. But this is not simply an argument resting on authority. These are the people whose job is to interpret scientific results, both their quality and relationship to the findings of others. And this, for me, is the clincher. Dr Paul Willis from the national scientific non-profit organisation RiAus in Adelaide (and ex-star of *Catalyst*) describes this as 'consilience' (a concept first offered by the aforementioned Rev. Whewel). This means that dozens of *different* lines of science measuring many separate aspects of nature (air chemistry, water acidification, ground and sea temperatures, bird migration, plant growth, many more) ALL point in the same direction. A murder case with twenty lines of evidence pointing towards 'who dunnit' would be accepted as offering more than twice the evidence required. That is the level of utterly convincing results that climate science has now reached. There is even a journal called *Nature Climate Change* with masses of the stuff. It is close to being the most established scientific phenomenon ever confirmed by research.

But it is never good enough for its critics. So, in my introduction to Lewandowski on that famous *Science Show*, I gave a somewhat provocative comparison. What would you think, I asked, if asbestos were offered as a good inhalant to cure asthma or the idea that paedophilia is good for kids? This was to suggest that the rejection of climate science holus-bolus is equivalent to such cavalier wrongheadedness.

Well, in a blink, the headline writers had their distortion conjured: I was comparing deniers to paedophiles. That this was rubbish was confirmed by the independent body whose job it is to investigate complaints to the ABC. But there I was on the front page of the *The Australian* newspaper, in its editorial, in its notes, and the subject of op-ed pieces by Maurice Newman (former chairman of ABC, the PM's business adviser), Janet Albrechtsen (former board member of ABC), Chris Kenny and, from the UK, no less, the egregious James Delingpole, who felt free to call me a green 'loon', never having met me.

Now, one piece in the newspaper may be unsettling, but six in a week is carpet-bombing. I found the experience over the top, hilarious in its overkill. You can see the standard editorial approach in most of the Murdoch-owned papers. You pick a victim for the week and then smear relentlessly. The ABC and SBS are favourite targets.

Those newspapers are also platforms for seeding doubts about climate science. And the result? Inertia in dealing with what is likely to be a monumental problem for all of us — well, not me, as I'm too old. In the USA only 40 per cent take the climate science seriously. Here the number is larger but the constituency for remedial action is small, mainly because people are uncertain about what to do and funding for renewable technologies has collapsed.

There is also the problem of the new media. I may slog, with my colleagues, to bring fair and independent comment on the subject of climate, but we are a minority. Not only is there a much larger commercial media, we now have internet sources that can back up any cockeyed belief someone may choose to entertain. Psychologists insist that the parading of contrary facts does little to change the mind of someone set in a prejudice. It is a matter of the person's values, not their information bank.

But this last insight offers a way through, at least with the non-ideologically driven citizen. Asking people to look around their district and to note whether anything has changed does open their eyes. This was done in an experiment run from Curtin University in Perth with fishers who lived for some of the year on the Abrolhos Islands west of Geraldton. They were unconvinced about the scientists' climate concerns. So a PhD student, Jenny Shaw, asked them to look around the islands to see whether any animals seemed different — more abundant, absent, or distressed — and whether tides or birds or plant life was altered. They took pictures, often rather beautiful ones, and the student arranged an exhibition and a book of the best ones.

Many of the observations matched the finding of scientists, who then added their comments to the exhibition display. As a result, most fishers now think quite differently. There *is* evidence, much of it their own. By examining their own familiar place of work or living in a way that reflected their own values, their attitudes, perhaps their minds had been changed.

> **Jenny Shaw:** We went over to each of the main island groups and asked them what changes had they seen and yes, they had seen a lot of environmental changes, including coral-bleaching for the first time. They've seen warmer waters, species that you would normally find further north, changes to corals and plants. And we also then asked them if they believed in climate change. There wasn't the same level of positive response. So we said to them could they take photos

of the changes that they'd seen and the things that they really loved about their industry and the islands. And we had lots of photographs, over 1000, and we actually put those into a large exhibition at the Geraldton Museum. So we wove the changes that they were dealing [in] with changes that the scientists were seeing as well.

One of the fishermen said it was like a light going on — it was just so obvious when they saw these photographs in a formal setting.

About seven years ago there was a huge drop in the number of rock lobster larvae along the coast. The state intervened to protect the stocks and dropped the catch by about half, so that had a huge impact on the number of fishers in the fishery and at the Abrolhos Islands certainly the number of fishers dropped by about half. So that really changed the community.

Abrolhos Island fishers:

> I've noticed two islands [are] gone — the Second Sister and Sandy Island. There used to be camps on Sandy Island, which are not there anymore. The actual island's disappeared. And Second Sister's been washed away.

> It opened the fishermen up to listening, being engaged in the [climate change] conversation. Now they are more aware and they are more open.

> [Initially when] asked about climate change I said it was a load of cobblers, whereas now you talk to people and I get different aspects of what other people think, so it's brought my thinking [around] too, that there is global warming.

This could be the way forward. It will take time. But, if this kind of citizen science helps with understanding their neighbourhood, who knows how quickly the science may come to be accepted?

A final thought on Australia's role following the 2015 COP Summit in Paris. The so called climate sceptics are very fond of claiming that this country is too small a player to make much difference either in contributing to cuts in global emissions or in developing green technologies. This argument is the standard excuse to do nothing about anything. Ultimately we all do relatively little and our lives are ephemeral, sure, but that lazy evasion won't do for reasons summarised so well in Michael Fullilove's superb 2015 Boyer lectures. This comes from the opening of his final talk:

> It is often said that Australia is a middle power. But there is nothing middling about Australia.
>
> There are two hundred-odd countries in the world. On every important measure except population, we rank in the top ten or twenty.
>
> Our economy is the twelfth largest in the world. Our people are the fifth richest.
>
> Australia is an old democracy and a free society. We are allied to the global leader and located in the most dynamic region in the world.
>
> Our diaspora is one million strong: our own world wide web of ideas and influence.
>
> We have a continent to ourselves. And we are fortunate enough to share it with the oldest continuing culture on Earth.
>
> Australia is not a middle power. Australia is a significant power with regional and global interests — and we should act like one.

7 Some Experiments in Broadcasting
Taking risks

I began some of the hoaxes in 1974, before *The Science Show* even took off. The first was in the then equivalent of *Ockham's Razor*, called *Insight*, where interviews were tolerated as well as scripted talks. We were coming up to number 500 and I thought someone really grand should be featured. Unfortunately, I found that most of the grandees — Mac Burnet, Gus Nossal, Mark Oliphant — had been on recently. And I could not think of someone on that scale to wheel in.

So I made him up: Sir Clarence Lovejoy. He had two Nobel Prizes, both for aspects of brain-flux theory (a made-up field) and he was pioneering ways to stimulate parts of the brain such as the lobus levitatis (centre of laughter stimulation) and the pons ponderosus (centre of pondering) to give us control of a subject's thinking.

The part was offered to Barry Humphries, who politely accepted, but was scotched by his agent who pointed out that Barry accepted everything before lunch but forgot it completely afterwards — those were his liquid days. So, instead, I found the ideal co-conspirator in Fred May, professor of Italian (and a little philosophy) at the University of Sydney. He was also a supremely gifted actor and an absolutely wicked wit.

The hoax was duly broadcast with, to my surprise, little opposition from my managers. The creative spirit was more buoyant back then. The result? A number of journalists asked to interview Sir Clarence and we both, Fred and I, appeared on a number of commercial TV programs explaining and then performing our ruse. It was fun and the public warmed to our frolic.

So what is the point of running hoaxes? It is a serious business, takes months of preparation, depends on selecting the very best performers and stands every chance of failing. Humour is deadly: if it is not working the audience knows immediately and you can see an abyss with no chance of salvation. But if it works you have listeners, decades afterwards, referring to The Fossil Beer Can or the premature broadcast of Princess Diana's first child, done as an outside broadcast with all the grovel and condescension of the old BBC royal genuflections on air.

It is essential that a piece of fun contains the strong scaffolding of a Big Idea. The Fossil Beer Can tried to reverse the evolutionary picture of human origins in Africa and the arrival in Australia as a misreading of indicators, rather like the Piltdown Skull had suggested (though a blatant fake) that British ancient history superseded all that 'foreign stuff' claiming to be older. Piltdown even had a fossil cricket bat, proving the priority of the white English sporting gent. I suggested that the beer can, surely an emblematic artefact for Auustralia, indicated that white humans were here in Australia first and one or two fell over, giving rise to the apes, as would happen if you were Pissed Person. Confusion about the direction of travel, not east from Africa but north from the Antipodes, came from the habit of this Australian person, *Homo micturans* (Pissed Person), of walking backwards. It was also significant, my script implied, that modern man had evolved, in this wide brown land, without women! Yet another challenge to evolutionary theory.

The genius proposing all this was a Professor Fraser, in the form of satirist John Clarke plus Scottish accent. Bob Hudson had played a tape on air of Clarke, in the guise of Fred Dagg, farmer, mustering stock and poultry on the paddock. I immediately recognised genius on stilts and phoned him in New Zealand.

Luckily he was coming to live in Australia in 1977 and so a number of *Science Show* conspiracies were conjured up and our golden period lasted for some years.

I wrote a script based on our new recruit in the Science Unit, Peter Hunt, who was a geologist. He simply filled a page with jargon and I added verbs. In between the nonsense we told the story of humans in Oz. No one objected to any implied insult to indigenous folk and we were careful not to stray into such explosive and *incorrect* implications, but we did, incidentally, criticise Aboriginal restrictions on the display of fossil material in museums. As it was, we prepared a demo of the 'fossil' beer can and had it displayed in the main foyer of the Australian Museum. The Temperance Society complained of our celebrating alcohol, irrespective of any importance of the fossil find.

The Diana indulgence, a hundred programs later, was based on the worry about having a head of state who was both Australian and royal. Alan Trounson's breakthrough development of frozen embryos for IVF implied you could achieve this by having Diana leave eggs in every governor general's fridge in the Commonwealth. The fertilised eggs would be both royal and local, which gave rise to the question: when did the essence of royalty actually enter the cells? Our answer, immaculately delivered by Dr Earle Hackett (the radio doctor who became chairman of the board of the ABC) and Tom Molomby (also a member of the board, now a leading barrister), was 'at the moment of orgasm'. This raised the question about royalty having orgasms — a quandary that arose much later as Diana and Charles embarked on relationships not observed in public since the cavalier adventures of Edward VIII.

Our program succeeded through the brilliant manipulation of sound effects by my colleagues, plus the impeccable impersonation by another genius, Tony Baldwin, of Richard Dimbleby, that old purveyor of Royal Unctuousness. Complaints were sent to the then chair of the ABC board, Dame Leonie

Kramer, who replied blithely that she thought the show was 'rather well done, just like a BBC extravaganza'. A flood of journalists' demands for a response from Buckingham Palace produced an official 'No comment'. This was quoted as a banner by the *Sunday Telegraph*. I still have the poster above my desk.

Other hoaxes came and went. We have had few recently because of absence of staff whose help once made it possible, and there are those concerns about a more unforgiving society where fun and games are not allowed. The commercial radio broadcasters who called the hospital where the Princess of Cambridge was lying gravid did something we would never have done: dupe the innocent and put them to air. All our co-conspirators were willing agents and we had clear signposts for the observant, therefore indicating that this was a leg-pull. But, just as Norman Swan and Chris Masters have discovered, with the perils of investigative journalism in these litigious times, don't bother. The court cases will last sometimes for decades. The spoilsports have won.

Other experiments also have mixed success. We have always gone with new technologies that work (we were among the first to try podcasts and even, for our sins, were the first to use talkback, or phone-ins, on ABC Radio back in 1974/75). We have tried production styles that challenge and we are willing, unlike our fellow science broadcasters in Canada and UK, to have a mixture of voices and contributions from all over everywhere, not just one presenter doing all the interviews.

But our experiment with the ground-breaking team from *Radiolab*, presented by a team from National Public Radio in New York and now breaking all podcast records around the world, met with immediate protest. The trouble was twofold. First, music was played incessantly under words, something older listeners hate; secondly phrases were repeated, inserted and spread in ways requiring concentration and a new way of listening. The payoff, for me and other fans, was storytelling on a grand scale and sound clearly evoking the nature of the topic.

Radiolab does it so well for two reasons: first, they have huge talent; and second, they have an enormous staff and may make nineteen substantial recordings for a show and then use only one. Their wastage rate is beyond extravagant. At *The Science Show*, with a team consisting of David Fisher and me (just to remind readers: no staff reporters or researchers) we would last a mere month with that approach.

But *Radiolab* amortises these extravagances through the size of its audience and even live shows on stage where material may be given extra outings. But you won't hear another *Radiolab* season on *The Science Show* — not because we are worried about protests from 'Outraged Toorak' but because you can get them more easily on-line, so we have done our bit to recognise a newcomer with flair, now it's up to you to seek them out as you desire. In the final chapters I shall consider how the future of broadcasting will treat shows like ours. Both *The Science Show* and *Radiolab* are an hour long. How does this square with the assumption, even the cliché, that attention spans are not much more than seven minutes? This is where technique comes in.

In the beginning I was encouraged to leave most interviews with people outside the ABC in their 'natural' rough-and-ready state. This was seen as allowing people to sound like themselves and not like ultra-smooth pundits speaking without hesitation or repetition. But, I protested, being broadcast is far from 'natural'; we need to assist these folk with every trick to make them sound good.

One disadvantage of this, as I was told by a shy scientist uneasy about an interview, 'I could never sound as good as those you have on!' 'Haven't you heard of editing?' I replied. 'You will shine like the rest, and that's a promise.'

My experimentation with such editing goes way beyond what you'd imagine. The first edit simply cuts out all the asides and irrelevances so as to leave only the essence of the topic being covered, apart from any juicy anecdotes or yarns (which are left in). The second edit removes verbal detritus — not just

Some Experiments in Broadcasting

the *umms* and *errs*, but the *Sos*... (why do so many academics always start sentences with 'So'?), *kind-ofs, you knows* and even *veries* ('very' is really quite meaningless — another punctuation word just stalling for time). The third edit — the polish — shortens breaths, removes unneeded clauses and even some of my questions, allowing the interviewee to flow.

How do I know all this effort is worthwhile? There is feedback from listeners, appreciating density instead of chatter; the ratings and certainly podcast numbers look good, and I am able to fit more stories into the show as items are shorter. Finally, podders often select one or two items instead of the whole show (damn them!), and, again, they prefer brevity and focus.

Broadcasting like this always has a tension between simple clarity, only words; and lively decoration, allowing the sound to add emotion, zip and extra meaning. I will not go for too much decoration if I think the item is self-sufficient. But if I think we could reach a level one-step up, I go for it.

When Mark Scott became CEO of the ABC he kept pointing out of the window saying that out there, in the wider population, was much more knowledge and ideas than we inside the ABC could ever muster. I objected, with some colleagues such as Geraldine Doogue, that it was our journalistic duty to sort that enormous knowingness, lest we become another Tower of Babel. This is one thing we do: report. We try to give shape and interpretation to what's 'out there', otherwise you're just inviting a cacophony. We are journalists. A public broadcaster should try to cover the field of human activity, not just the glam bits, and make the unexpected of interest.

Having said that, we in science radio probably give more exposure to 'outside' people than any other part of the ABC. For a start, *Ockham's Razor* has, for more than thirty years, broadcast talks by members of the informed public every week. We kicked off with Professor Peter Farrell slamming the Australian university system; he is now head of ResMed, the multinational medical technology company. On *The Science*

Show, we have had not only short talks but entire series from writers such as Peter Mason, Tony Barnett, Jane Goodall (University of Western Sydney), Mark Dodgson, Derek Denton, Dr Jonica Newby and many more.

Most who ask are told 'yes'. There are no ideological constraints or restriction on topics, only ones of scientific integrity and good use of language. Of the former, Peter Farrell is, again, a good example. As we approached the thirtieth anniversary of *Ockham's Razor*, I invited him to celebrate with another talk. I suggested innovation as he was a fine example of a successful innovator and was writing about the topic for possible publication. Peter replied that he would, instead, prefer to write about climate science being a hoax. I said to him: 'How do I square that with my being a science reporter reliant on evidence?' On the one hand I have tens of thousands of first-class climate scientists doing research in the field saying we have a problem. On the other there is Peter Farrell, an expert on snoring, saying we haven't. Is this not a manifestation of pure ideology, Peter being of impeccable right-wing pedigree? Are we not back to the flakiness detailed in Chapter One, when anyone could say anything they felt like whether it added up or not? There is the added editorial problem about risk. Some subjects (HIV as a cause of AIDS, many vaccine reports) may ultimately endanger lives if we allow a commentator to contradict the firm facts. This was perceived to be the problem with Maryanne Demasi's report on statins in *Catalyst*. Going off the drugs could have risked lives, some said (like Norman Swan, vehemently). It would be similarly irresponsible if I told people to ingest sheep drench as a cure for heart disease.

One experiment I tried, with heartbreaking success, was with a friend, a doctor, who was suddenly diagnosed with deadly mesothelioma from an asbestos shower that fell on him in his home when a ceiling collapsed. He knew his fate. He seemed sanguine about it. I asked him whether it would help to keep a diary which he could broadcast as time passed. He was willing and said, after the first couple of episodes, that it gave him a lift — made the death sentence just a little more endurable.

This is part of one extract.

Dr Jim Holmes: I am not a religious person and regard spirituality in socio-biological terms. Nevertheless, whether or not certain phenomena are regarded as spiritual or the result of cerebral activity, the result can be the same. Either way, meditation, positive imagery and attitude can result in improved outcomes, at least in quality of life.

It seems that the mesothelioma is advancing as one might expect. The chest X-ray I had three weeks ago showed less pleural thickening than six months ago, before chemotherapy, but more infiltration into the lungs. I'm not sure how much of that is due to infection and how much due to the meso. I must ask my oncologist next time I see him.

The deterioration in my breathing before the X-ray was quite rapid and I therefore assumed it was secondary to infection. However, after a slight initial improvement, there have been no further gains and my exercise tolerance remains considerably less than before.

Nevertheless, I took myself to Bellingen [north coast of New South Wales] for Camp Creative where I participated in a choral workshop for the sixth year running. Mind you, there was no running anywhere this year; I could only stroll the 50 metres from my car to the venue at a snail's pace and even then had to sit for a while to recover my breath before singing anything.

I arrived with my wheelchair and oxygen concentrator, a bar fridge-size machine that pumps air and increases the percentage of oxygen it contains. However, I managed without using either contraption, but I did use my shower stool without which I simply became too short of breath for the exertion of standing to wash.

I had also starting taking steroids that week to reduce the inflammation in my lungs and to assist my breathing. I took dexamethasone, 2 milligrams, morning and lunch. I was ambivalent about taking this due to the potential side effects of chronic usage. These include osteoporosis, diabetes, hypertension, gastric ulceration, and immunosuppression, which I certainly did not want. I therefore cut back on the dose after a few days. Big mistake.

Within two days my breathing had deteriorated further and I had to resume the original dose. By the end of the week my group had assembled a repertoire of about ten pieces, some of which we performed for the camp's final concert. I would have been barely visible to the audience because I was mostly seated at the back of the standing choir, supporting the base line with hopefully sonorous tones punctuated by breaths more frequent than desirable.

I think I'll stay put for a while, a couple of weeks at least. The poor garden needs a lot of attention and I can at least put my chair among the vegies and sing to them.

Dr Jim Holmes was a friend. I met him and his family in 1975 as I moved in to Balmain and started *The Science Show*. He was unusually gentle and highly intelligent. He practised as a psychiatrist both in Sydney and, later, in Kempsey, New South Wales. He had a long black beard and was fond of tending his country garden full of vegetables and fruit trees. He was not a toiler in old buildings or with a history of rough work where he could expect to be exposed to asbestos. His diagnosis was as unexpected as you could possibly imagine, and when it comes you have no choice but to accept it. Like rain and earthquakes: that's the deal. Jim's broadcasts, calm and straightforward as they were, showed how to cope. We miss him.

Ultimately, each week is an experiment. How does material fit together in a way that people will enjoy and find engaging? You think you know when you are doing it, but when you listen off air, as I always do, the live program gives me a different response, sometimes a good one, sometimes bad. Is there too much of me? Are stories crowded together too closely so you get no respite from the torrent of words? Does the format suit the time? Now *The Science Show* is broadcast not only at its usual slot of midday on Saturday, but also at 4 a.m. on Monday and 9 p.m. on Thursday, I have given up trying to think of that aspect of varied listening. Weird scheduling.

I have also nearly given up on the feedback so essential from colleagues. A few are terrific and say something useful, beyond the two sentences indicating that they may have caught two minutes. In times of yore, outside any formal editorial meeting, colleagues would mention production ideas I may have missed or not bothered with. In the old days when we always went live, I'd get a phone call in the studio from the doyen of current affair presenters, Paul Murphy (former host of ABC Radio's *PM*) with some delightful perceptions I could find useful. Nowadays, with everyone Araldited to their screens and impossible schedules, you're lucky to get a couple of words slung over a shoulder as you pass in the corridor. What do you do in a world that my newspaper today tells me allows half an hour a day for women to interact socially and barely twenty minutes for men? Are there no personal conversations, only meetings?

In the end that's what we do: broadcast conversations. Whatever the new technology, however slick the production, we are left with what was there in the beginning: people talking to people. Our job is to make it compelling, even unmissable. Simple as that.

8 Stars

Our stellar performers

Our first real star was Peter Mason. Peter was foundation professor of maths and physics at Macquarie University, New South Wales. I cannot recall where I met him but he was like a number of fellows you know who are around. You see them here and there, start saying hello, have a brief conversation and next, you're almost friends, even though there have been no invitations to home or pub.

Peter was warm, immensely enthusiastic and positive. He would share miserable memories of despots and cruelties and reflections on injustice in this wicked world but they would always be followed by remedial optimism, without a trace of sentimentality or utopianism, showing how the Promised Land, in a secular sense, could be achieved in our lifetime. He was no Pollyanna. He really believed in the essential goodness of people.

One day I asked him to do a book review. He agreed before I'd even finished my sentence. The following week it was ready. He came into the studio and read it in a sing-song voice. It was pleasant but unremarkable, but good enough. I put it to air. Then I sent him a few notes on what was wrong with it and how to improve. The following day he had absorbed all the advice and wanted to go again. He did. And it was three times as good.

Peter listened and learned like no one I'd ever met. Then he announced he'd written a three-part series on the history of navigation. Oh dear, I thought, why not a series on the nature of reflux or three half-hours about the pencil? Then I saw the script. It required actors, music, a couple of famous figures from overseas and ... what could I do? Saying 'no way' when he had actually written the thing would be cruel. Maybe I could just do a rushed job and bury it in the summer season.

This is what happened. We knocked it off in the studio in record time, bunged in a couple of out-of-work actors (they turned out to be stars themselves) and chopped the parts together. On went the series, 'Genesis to Jupiter', as schools broke up and Aussies repaired to the beach for a couple of months.

The response to the series was unprecedented. In those days (no emails) letters were written. We received kilos of them, for weeks. A publisher wanted to make a book of Peter's script. The phones rang demanding repeats. What was happening?

Peter had hit a radio jackpot. He was a born storyteller and going back to the Bible and the ancient Greeks was no impediment. Our audience loved the history, the way the stories linked together, how the dove in Genesis, let loose by Noah to indicate a distant shore, had led through history via maths, tunnel-drilling and astronomers, to NASA and its *Voyager* spacecraft going to Jupiter and doing so indirectly, looping, as the planets moved. Navigating space–time.

Next, Peter wrote us a series on light ('The Light Fantastic'); one on the Nazis and the role of iron in the body as well as war ('Blood and Iron'), and another on rubber. Yes — a history of rubber from the Amazon to Southeast Asia, from Caruso to Irish rebel songs. Eclectic in the extreme. These series were loved by everyone from Bronwyn Bishop to Barry Jones. Books were printed, sold out, printed again. It was Peter's soft, smooth voice, command of the topic and incredible hope of progress that made everything work.

Then, tragedy. Peter was working with a huge team to make the ANZAAS Congress (The Australia and New Zealand Association for the Advancement of Science — equivalent of the AAAS, but now gone) the best success it could be, without interference from the suits, and so open doors for young people and the public. One of his helpers was Patrick Moore, star of *The Sky At Night* on the BBC for fifty years.

I met Peter just before the big day, the opening. And he'd forgotten his home phone number, one he'd had for decades. This, from a maths prodigy. Overnight, a brain tumour had eradicated his numeracy. His response?: 'Oh well, having done so well with my left hemisphere, I shall now concentrate on my right one and do more arts.' Even at the end he was trying to find the positive side. He died bereft, having been saved too often. Even his membership of a society to grant an easier exit did not convince doctors to leave him alone — saved to suffer. It was the only time I saw his smiles disappear. People still ask about his series.

* * *

And so to some other scientists who could fill a stadium, just like Brian Cox and Dr Karl with Adam Spencer can do today. I have chosen a few for this chapter to give a feeling of their range and articulacy. It is a set of extracts maybe to dip into. Select and savour.

Why are people now so willing to discount the stars of yesteryear? They are not just Dead White Men. They have not been eradicated by some mindless cleansing force as demographics have changed. Once, on a long plane ride across America, I could no longer read and I'd seen the film, so I asked my neighbours in the next seats, 'Who is the most famous scientist in the USA?' Nobody could think of any scientists at all, let alone famous ones. I had to go back four rows before a soldier said brightly: 'Carl Sagan.'

'Well done,' I remarked, 'I'm afraid he's dead.' But he was right — Sagan was one of the most famous scientists of the twentieth century. And others? When I started *The Science Show*, and for the next twenty-five years, Professor Paul Ehrlich of Stanford was a very big deal. Nowadays, people tell me, almost proudly, '...never heard of him'! It's as if the demographic wants to bury the scientific equivalents of Max Bygraves, Englebert Humperdinck, or Esther Slag (OK, I made that one up). History as fashion!

Yet all our stars have made significant contributions to the major ideas of the twentieth century. Test yourself with these names, then look them up if you're left floundering: Stephen Jay Gould, Margaret Mead, Mark Oliphant, Mac Burnet, Jack Eccles, Isaac Asimov, Oliver Sacks, Douglas Adams, Howard Florey, Dorothy Hill, Sally Ride, Ted Ringwood, Jared Diamond, Valentina Tereshkova and Julius Sumner Miller.

All have been on *The Science Show*, some at length. Paleontologist and evolutionary biologist, Stephen Jay Gould would, in the 1980s, fill a large hall. I met him at the New York Hilton in 1988 when he was still fighting a kind of mesothelioma, from which most people die quickly. Gould survived, but was so thin I offered to carry his briefcase to the large auditorium. It was packed; so much so that the staff began to move the partitions to accommodate the overflow into the adjacent ballroom. Gould had already begun to speak and *did his nut*! There is no other way to put it. He was a singularly determined man with a rigid routine. He spoke, you stayed still; any noise or flashlights and he became incandescent.

But he was always mild with me. I'd talk to him about his snails, ones he'd collected on islands to see how they adapted quickly to changing vegetation and colours. We discussed his views on punctuated equilibria, how evolution may go in spurts as the environment changes, preferring to stay stable if given a chance. Our last encounter, when he was president of AAAS, was to discuss God and science. He insisted they are two separate 'magisteria' or realms, capable of coexistence, but unlikely to come to an accommodation because their assumptions and language are so different. Gould as diplomat — a rare manifestation.

Gould could be very funny and ultra-serious. I saw him once give a 90-minute lecture on the history of baseball scores. Not a joke emerged, nor much of an explanation about why this subject was being featured at a major science conference. Yes, the signs of Asperger's were everywhere. His books are still thrilling to read. Most are made up of essays written for a natural history magazine in the USA. Among the yarns, you will be reminded that Darwin and his captain, FitzRoy, were barely twenty-three when they set out on their five-year voyage; and that Darwin was not recruited to be the ship's natural historian, but instead to talk to the captain as no other member of the crew was in the proper class (upper middle or aristo) to be allowed to sit at the captain's table. So, for five years, that was the arrangement. Gould read and Gould knew. His ideas are not extinct.

By the way, did you know that Captain FitzRoy was the father of weather forecasting? He applied science to the discipline and helped make it reliable. But when his work was attacked (forecasting is a tricky business) the man who had steered Charles Darwin around the world for five years killed himself.

Here's some Gould:

> Evolution and Darwinism is a set of notions of enormous beauty and power and importance, both directly for our lives and in the abstract. It has been a series of notions long debated by folks in whose shadow we can scarcely walk.
>
> My claim about the nature and status of Darwinism is that we're beginning to see the coalescence of a new and quite different theory from that which represented the strict Darwinism of the so-called modern synthesis. A new and different theory with a Darwinian core is being forged, an exciting and fruitful theory in the very best sense that first of all it is quite different, in many important respects, from what was the standard take.

Darwin always made a clear distinction between the two fundamental things he tried to do. One [was] to establish the fact of evolution, in which he was abundantly successful, [and] that's why he lies at the feet of Isaac Newton down there in Westminster Abbey. Secondly, however, he tried to propose a theory to explain the mechanics of evolutionary change, the theory of natural selection, which was by no means a majority viewpoint upon his death in the early 1880s. And that's what we're talking about, his theory of natural selection.

What is the theory of natural selection? How can it be epitomised? I think the minimal requirements can be reduced to two fundamental propositions. The first is the claim not just that natural selection operates, because all non-Darwinians were happy to avail themselves of that opinion, but that natural selection is the creative force of evolutionary change. And that's somewhat paradoxical because natural selection doesn't make anything, so in what sense of the vernacular word 'creativity' could it be a creative force? It can't be just a negative force.

I think the most brilliant thing that Darwin did was to recognise that in order to have evolution work in such a way that natural selection could be seen as a creative force, that variation would have to be structured in a very specific sort of way, and in a most brilliant manner, knowing nothing about the actual mechanics of hereditary change, he correctly predicted the fundamental characteristics of the nature of variation by knowing what would be necessary in order to allow natural selection to be creative. And those are three main properties. Variation has to be copious because natural selection doesn't make anything so there has to be a lot of raw material to choose among. It has to be small in extent because if variation is so big that you can get from here to there in one step then you've gotten there by the force of variation and all natural selection does is eliminate the bad ones. Thirdly, variation has to be undirected because if variation comes packaged preferentially towards the newly adaptive form then random mortality that produced the trend doesn't need natural selection at all.

And Darwin has a very definite and very interesting answer, which is the heart of his radicalism as a thinker. Darwin answers very simply; natural selection works on organisms. Natural selection is the struggle among organisms, among organisms' bodies for reproductive success. Why is that so important? Why is that at the heart of Darwinism? The answer is: think of what he's trying to do, think of the heart of the radical claim.

That's a radical argument. All the higher order harmony, which used to be seen as the source of evidence of God's benevolence, is epiphenomenal upon the struggle of something lower for personal gain, and that's all. Now, if that sounds to you like Adam Smith's economics translated into nature, it's no accident because one of the most interesting conclusions of recent historical research has shown the tie of Darwin's development of the theory of natural selection in 1838 to his interest in the particular view on individuality in the thought of Adam Smith and the Scottish economists. It is the same argument. Namely, that if you want the most rationally ordered economy you don't get all the smart people, sit them around a table and pass laws for higher order harmony of economy, as Adam Smith argues, you do something that seems paradoxically opposite; you let individuals struggle for profit. And the ones who do it well will knock out the others and balance each other out and you get the maximal system through the action of the invisible hand.

The analogue of profit in nature is reproductive success.

Stephen was a giant among intellectuals, but I doubt he saved many lives. Howard Florey did though. The comparison I rather enjoy is that Tom Cruise may save around twenty lives in *Mission Impossible* but Florey saved *hundreds of millions* through his brilliant work on penicillin. Ian Frazer's work on the vaccine Gardasil must have saved hundreds of thousands, rising soon to millions. These stars are giants, not equivalent to faded pop stars or movie actors whose names you can't remember.

Talking of Florey: the foundation director of the renowned Florey Institute in Melbourne is Professor Derek Denton. You'd be forgiven for mentioning him as merely a partner, as his wife is Dame Maggie Scott, founder of the Australian Ballet School, which she ran for forty years. Both are AC, the highest honour on the Australian list. But Dick (why don't they call him Derek?) is still publishing front-line scientific papers at past ninety years of age. His great work is the book *Hunger for Salt*, which explains how our brains are geared to seek out salt and water. Thirst is a driving force in all vertebrates. And Dick's latest papers have shown, as a result of impeccable lab trials, that our addiction to narcotics has partly invaded the same brain areas as this most powerful of drives, seeking out water to slake our thirst. No wonder those addictions to drugs are so strong. Having found the seat of the problem, could there be a new route to find cures?

Some time ago Derek Denton wrote a series for *The Science Show* on the importance of brain research.

> The ultimate challenge to human ingenuity is the unravelling of the nature of consciousness, the awareness of being alive. It's a major pre-occupation of the enquiring human spirit. It's the scientific challenge with the greatest implication for the human future. Einstein and Freud, in their renowned correspondence, identified it as the most pressing task for humanity.
>
> The challenge is not an esoteric or academic issue.
>
> The examination of the issues has been anything but the prerogative of medical biologists. The questions of

consciousness and mind have preoccupied philosophers, writers, poets and natural scientists for millennia, from the early Oriental and Greek philosophers onwards.

For many human beings there is an overt delight in the feeling of being a part of the physical world, involving as it does the awe of the cosmos and the physical pleasure of the forests, rivers and oceans.

The phenomenon of consciousness has arisen progressively in the course of the evolution of animal life because its emergence, elaboration and refinement has conferred great survival advantage on the species. The advantage lies in the creature being able to exercise options by images in the mind, be they the most rudimentary or the most elaborate. The animal may examine the possible outcome of its actions. It can choose a course and in so doing may meld its instinctive memory, the legacy of the past, with such experience as it has already had in the course of its own life.

Henry James, the distinguished American writer, expressed the view concerning the options as mind being a theatre of simultaneous possibilities, the selection of some and suppression of the rest, whereas William James, eminent psychologist and philosopher, and brother of Henry, suggested consciousness is what you might expect in a nervous system grown too large to steer itself.

We are still arguing about the nature of consciousness. Both Dick Denton and brain scientist Susan Greenfield agree that it has evolved and so you would expect other creatures to have versions of being conscious, the way a dimmer light can get brighter and brighter, until one day, in humans, it is full on, at its peak so far.

Tony Barnett was not exactly a star. He was a leading international intellectual from Britain whom I met at the ANU where he was professor of zoology. Tony was the major critic of Desmond Morris and his 'Naked Apery', in which Morris would isolate aspects of our mammalian past to show how we are locked in to certain behaviours. Men react to breasts because they are reminded of buttocks and therefore of intercourse. Women display 'nest building' (please, Jonica, not *more* cushions!) because they are wired to become mothers. Men need to leave the nest, or village, to go to work or explore because they are conditioned by brains fashioned during our hunting past. Now, all this may be true to some extent, but Tony's mission was to assure the world that it's not *inevitable*. The point about humans with big brains, he insisted, is that we can override our jungle tendencies and erect any social system we choose. Some have been built on egalitarianism and tolerance, others on savagery and evil. Our ape ancestry determines little.

Evolutionary change is piecemeal. It consists of bit-by-bit modifications of what's already available so imperfections shouldn't surprise us. They've been called the results of tinkering. Peter Medawar (Nobel Prize in Physiology or Medicine 1960) has even said that 'Evolution is the story of waste, makeshift, compromise and blunder'.

Perhaps it's an overstatement. But blunders or not, evolution is still going on. It's a continuous process of adjustment to circumstances. It can never be complete because the conditions organisms live in are always changing. So if we want to explain ourselves biologically, we are faced with major obstacles. But the truth is often obscured. You sometimes hear of genes for various human attitudes: a gene for altruism, a gene for gratitude, a gene for sympathy, a gene for guilt, a gene for establishing reciprocal relationships, a gene for learning about establishing reciprocal relationships, a gene for learning about altruistic and cheating tendencies in others.

It sounds a bit like one of those lists of cranial bumps made by phrenologists. The scientific content is the same. Nil. Those genes are inventions. They are an aspect of what has been called:

The great aberration of thought that runs through sociobiology — the belief that human social structures, the rise and fall of nations, and all the nuances of human behaviour are genetically determined.

That was Peter Medawar again.

Modern genetics is a great achievement but it does not identify genes for moral worth or for general intelligence or special skills or for poverty or criminality or for liking dogs. A person isn't bright or stupid, healthy or unhealthy, sane or insane, like flowers coloured red or white. Yet popularisers persist on saying that they have a method of interpreting human society based on evolutionary theory and genetics. If so, they should have something instructive to say about one of the major social transformations of our time.

In 1970, in the USA, women made up 16 per cent of managers and executives; now the figure has passed 43 per cent. In recent times (1998) 35 per cent of university teachers were women and even in science 22 per cent. Within living memory, figures like these would have seemed to many people beyond belief.

It's also worth noting these are not genetical changes, they are not due to a sudden outbreak of mutant genes for feminism. Not even the most fanatical gene-worshippers have suggested that — as far as I know.

Rapid and drastic non-genetical social change is typical of humanity and it has no counterpart in other species.

Yet popularisers still go on offering a lot about sex and evolution and genes.

If Peter Mason had a history of rubber, Dr Lindsay Sharp and Louise Crossley had a history of salt. This was, again, something that spanned physiology, Roman civilisation (salaries), fishing, deserts and much more. Lindsay was, at the time, founding director of the Powerhouse Museum in Sydney, and Cambridge-trained biologist Dr Crossley was his deputy. She died in 2015.

Lindsay went on to run the Royal Ontario Museum in Toronto and then the Science Museum in London.

Louise Crossley: It is early in the morning of April 6 1930 and dawn gives promise of a blazing hot day to come. The long rollers of the Arabian Sea break into surf on the beach near Dandi, a remote village lost in the desolate mud flats of India's northwest coast. From a makeshift hut, a small brown figure, clad only in a loincloth, emerges. As the low sunlight strikes across the waves he takes a brief dip in the shallows then, surrounded by figures in loose, home-spun robes, he stoops down and picks up a handful of salt from the tideline.

He is Mahatma Gandhi. And from this simple gesture sprang a civil disobedience campaign that was to lead India to freedom after nearly 200 years of the British Raj.

He would overthrow the greatest empire on Earth with a pinch of salt. Every year the tax on salt yielded a staggering £25 million to the British government and it affected all Indians, rich and poor alike. A declaration of war against the salt tax gave Gandhi a magically simple formula for mass civil disobedience. He would march to the sea, pick up a few grains of salt, and sell it. His small handful fetched the equivalent of $200 at auction.

By breaking the law forbidding the collection, manufacture, sale or use of any salt except under government licence, he would become liable to arrest. But his gesture would be the signal for every Indian villager to follow suit and could the government arrest 300 million people, or three million, or even 300,000? It was a brilliant political challenge with a dramatic and symbolic force that no one could mistake, least of all the government.

Lindsay Sharp: But why so much blood, sweat and tears over salt? The Salt March (led by Gandhi) is only the most dramatic example of the enormous significance that salt has had in virtually every human culture on every continent.

There's a fundamental physiological fact at the root of all salt symbolism. This is simply that blood, sweat and tears, and every other bodily fluid contain salt. And so if we excrete it in any of these forms we must also replace it in the food we eat. An adult human being, in fact, contains the equivalent of about 300 grams of sodium chloride and this saltiness is a legacy of our evolutionary origins. The first living cells developed in the saline environment of the oceans, and even though we've come a long way since then — up out of the sea, into the trees and then down to earth — our cells still operate as if they were back in the briny. This means that our survival depends upon maintaining the right concentration of salts in our tissues to within a quite narrow tolerance.

* * *

Oliver Sacks did not do a scripted series but instead plenty of long interviews and speeches that I recorded with him at large theatres. His shyness is overwhelming and unfeigned. On the first outing, at Sydney University's Seymour Centre, he would not leave the dressing room to come on stage. Many minutes passed and the audience stayed calm — just! So I went upstairs, took his hand, said I would be on stage to fill any hesitations, and then I led him down, trembling, to his fate.

As he appeared a huge round of applause greeted him. Oliver Sacks then talked delightfully for two hours. Getting him *off* stage was nearly as hard as getting him on.

Oliver Sacks: We don't have any way at the moment of studying the neurology of the imagination. I hope there will be a way in the next millennium. But one can study some of the lower levels in the nervous system, and this may be very valuable if there are two analogies between what goes on, say, at lower levels of perception, visual perception say, and imagination.

We are not given the visual world as a whole; there are something like forty or fifty different systems in the brain, which are involved in the breakdown, and reconstruction in the analysis and recreation of a visual world.

I've written at some length about an artist who suddenly became totally colour blind as a result of a knock out of one of these elements. And not only did he lose the ability to perceive colour, he lost the ability to imagine colour. He lost the ability to remember colour, he lost the ability to dream in colour. It's become very clear that the same areas, or similar areas of the brain, are involved in perception and imagination and memory. And to some extent these all go together.

These are peculiar syndromes in which particular elements are knocked out. But there's also an interesting condition called visual agnosia, when you can see perfectly well but you don't know what you're seeing, you can make no sense of it. This can sometimes happen as a result of brain damage. It happened in *The Man Who Mistook His Wife For a Hat*, as a result of disease affecting the visual parts of the brain. And he found himself in a world of colours, shapes, movements, with less and less meaning, which allowed the possibility of such a ludicrous mistake as mistaking one's wife for a hat or a hydrant as a child, or something like this.

And I want, if I may, to read you a wonderful passage in which Mozart describes, or tries to describe some of what happened to him as he composes. He says, 'When I feel well and in a good humour, or when I am taking a drive, or walking after a good meal, or in the night when I can not sleep, thoughts crowded my mind as easily as you could wish. Whence and how do they come? I do not know and I have nothing to do with it. Those which please me, I keep in my head, and hum, at least others have told me that I do so.

'Once I have my theme, another melody comes, linking itself to the first one, in accordance with the needs of the composition as a whole. The counterpoint, the part of each instrument and all these melodic fragments, at last produce the entire work. Then my soul is on fire with inspiration. The work grows, I keep expanding it, conceiving it more and more clearly, until I have the entire composition finished in my head though it may be long.' And at that point he says: 'Then my mind seizes it, as I glance my eye at a beautiful picture or handsome youth, it has not come to me successively, but it's in its entirety that my imagination lets me hear it.'

I think this is a marvellous description of the fragments coagulating to form the whole.

But not everyone is a Mozart.

When we had our *Science Show* twentieth-anniversary bash at the ABC in Ultimo, I was in mid-speech when a familiar tall figure emerged from the lift. I hesitated in surprise and then said: 'A most surprising guest, folks. He lives in New York but is with us today; please welcome Dr Oliver Sacks!' A friend had seen him somewhere and mentioned our anniversary. Out of kindness and friendship he just came. That's Sacks. I like to boast that neither of us had appeared in *Awakenings*, the movie, based on his book; the guy playing Sacks was that other Robin Williams and the Father Christmas in the streets of New York played by Sacks was cut out of the finished film.

Ollie, as he signed his letters to me, died in 2015. I put together a special *Science Show* drawn from all of our many encounters. As I did so I was struck once more by his shining intelligence and his almost overwhelming learning. I was also thrilled by some photographs from his youth displayed in his final memoir showing a muscled stud on a motorbike as remote from the slightly fumbling donnish figure of later years as you could imagine. Ollie loved mischief as much as he cherished scholarship. Such a splendid spirit!

Jane Goodall is a name to conjure with. She is the superstar who showed that chimps use tools. Louis Leakey recruited her despite her lack of formal qualifications. What a coup! She has travelled the world ever since, trying to save the lands where apes used to flourish. Will they become extinct because of our rapacious invasions?

> I see the chimpanzees and the other great apes as being like ambassadors, reaching out to us across this supposed gap and they need our help because they are becoming extinct.
>
> When I began my research in 1960 there were probably about a million chimps, 100 years ago nearly two million perhaps, based on the extent of suitable forest habitat across Africa. And, as you all know, forest habitats are decreasing with terrifying rapidity all around the world. Even the precious Gombe chimpanzees [of Tanzania] are threatened. When I arrived, forest habitat stretched right along the eastern shore of Lake Tanganyika, which is where the Gombe National Park is, and you could climb up mountains that rise up out of the lake and look eastward to the chimpanzee habitat. There were some villages but, by and large, it was all places where chimpanzees could live.
>
> Fifteen years ago I flew over Gombe in a small aircraft and I was absolutely shattered by the extent of deforestation outside the tiny 30-square-mile national park. It's the smallest in Tanzania. Within the park things are much the same as they always were, but whereas when I began there were 150 chimpanzees in three communities, about fifty each, today it's only the central main study community that remains a stable fifty. On each side the communities have shrunk because they used to move outside the boundaries of the national park and, as I saw from my plane, the trees have gone.
>
> It's very clear there are far more people living in this area than the land can possibly support.
>
> As I flew over, the question suddenly jumped at me: how can we even think about saving these so-famous

chimpanzees if the people living in the area are having a real struggle to survive? That led to our program Take Care, and it's a program to improve the lives of the people living around Gombe in an environmentally sustainable way.

Around many of the villages, where they were surrounded by bare slopes with just seemingly dead tree stumps where the women were hacking away to get firewood, now they're leaving the tree stumps, and within five years they can regenerate into a 20- or 30-foot tree. Take Care forests are springing up around some of the villages and that gives the one opportunity for the Gombe chimps to find leafy corridors so that once again, as they used to, they can move out of the park to make contact with other remnant populations, which they will need to do to increase the gene pool, or maybe some of the other remnants will visit them.

Meanwhile, across the rest of Africa, the situation is even grimmer. In West Africa, chimps are highly endangered and they were subjected for years to the live animal trade, where you shoot the mother to get the baby for medical research, for zoos, for circuses and so forth. In central Africa there was also, of course, some of the live animal trade, but in the great Congo Basin you have the last significant populations of the African great apes — the chimpanzees, gorillas and bonobos. And in the Congo Basin these great apes, along with the other creatures living in these forests, are under dire threat from the so-called bush-meat trade. The bush-meat trade is different from subsistence hunting where people hunt in order to survive. The bush-meat trade is commercial; it is the commercial hunting of wild animals for food, made possible by the foreign logging companies that have gone in and made roads deep into the heart of the last great rainforests of central Africa.

Even if they practise sustainable logging, the forest life is threatened because of the roads, so that hunters can go in from the towns, camp at the end of the road, shoot everything — elephants, gorillas, chimps, bonobos, monkeys, antelopes, pigs, even birds and bats — smoke the meat and they now have transport to get it into the towns where the urban elite will pay more for this bush-meat than they will for chicken or goat, because there is no culture in these parts of Africa for domesticating animals. It's totally unsustainable.

One of the by-products of the bush-meat trade is that whereas in the old days a hunter would not dream of shooting a mother with a baby — I mean, they're thinking of the future — but today, because they're thinking of money and dollar signs, mothers will be shot to sell for food, but you can't sell a baby chimpanzee for food, there isn't any meat on it. So, you'll see sometimes the pathetic sight of a mother's butchered body being sold for meat and beside her the sad little infant is tethered. Maybe somebody will buy it to bring custom to a hotel or bar, maybe an ex-pat will feel sorry for it. It's illegal now in all the range countries for endangered species such as chimpanzees and gorillas and bonobos to be shot without a licence, but there's very little enforcement of things like that.

There is hope for the future but not if we sit back complacently, not if we allow ourselves to feel helpless because, yes, the problems are huge around the world and there's nothing we can do, as an individual, to do anything about them. But each one of us can make those decisions every day that, ultimately, is going to cause social and environmental change and that we need to move out of this materialistic rut that we've fallen into, this ultimate consumer society, and move in the way of developing our psychological, our spiritual beings, if you like, so that we can truly find the space that I think the human being should attain in the rest of the animal kingdom.

* * *

Jared Diamond knows lots about Darwin and evolution and has written books about our own flirtation with extinction: *Guns, Germs and Steel* and *Collapse*. There is no doubt that many of our civilisations have crumbled, often in misery and strife. I am always reminded of the appalling figure that more than 90 per cent of the human population of the Americas was wiped out as the Europeans invaded. We can be a truly evil bunch. But what makes this so common? Is it the burning-deck prescription of the economist Schumpeter, who suggested that wars, plagues and other crises are necessary to clear the ground for progress? If we maintained our pleasant, old-fashioned communities we'd be stuck in some ancient past, convivial maybe, but without the life-saving benefits of sewers, public health, antibiotics and transport.

> **Jared Diamond:** When I started trying to write about the excitement of science for the general public, I quickly encountered an occupational hazard of the academic who would like to communicate to a broad public, and that occupational hazard is to encounter the journalist who says to me something like the following — and it happens again and again — 'Mr Diamond, I know that you've devoted the last seven years of your life to reading those hundreds or thousands of books and articles and condensing the history of everybody into this 420-page book [*Guns, Germs and Steel*], but, Mr Diamond, please remember, our TV viewers and our radio listeners and our newspaper readers are busy people. Would you please summarise your book and all of history in one sentence for our busy readers and listeners?' And I've learned to summarise everything in one sentence. The one sentence would go something as follows: 'The broadest pattern of history, namely the differences between human societies on different continents seems to me to be attributable to differences among continental environments and not to biological differences among people themselves.' And once

you've given the journalists that one sentence you can usually slip in a second sentence with many clauses. 'In particular, the availability of wild plant and animal species suitable for domestication and the ease with which those species could spread without encountering unsuitable climates, contributed decisively to the varying rates of rise of agriculture and herding, which in turn contributed decisively to the varying rates of rise of human population numbers, population densities and food surpluses, which in turn contributed decisively to varying rates of rise in epidemic infectious diseases, riding, technology and political organisation.'

Professor Steve Pinker of Harvard compares the life spans and conflict of yesteryear and finds us much more peaceful and long-lived, despite the bad news in the papers and on TV of murder and mayhem. The following passage is taken from an address to the RSA in London broadcast on the RN *Counterpoint* program in December 2011.

Steve Pinker: Instead of lamenting why is there war perhaps we should ask why is there peace? Rather than what are we doing wrong perhaps we should ask what have we been doing right? Because we have been doing something right and it sure would be good to find out what exactly it is. Also I think the decline of violence calls for a re-assessment of modernity, of the centuries-long trend that has eroded family, tribe, tradition and religion and allowed them to be superseded by individualism, cosmopolitanism, reason and science.

Everyone acknowledges that modernity has brought us many gifts, such as longer and healthier lives, less ignorance and superstition, and richer experiences, but there's always been a culture of nostalgia and romanticism that has questioned the price. Is it worth it if we have to live with the spectre of terrorism, genocide, world wars and nuclear weapons?

Despite impressions, the long-term trend, though certainly halting and incomplete, is that violence of all kinds is decreasing and I believe that calls for a rehabilitation of the concept of modernity and progress. It's a cause for gratitude for the institutions of civilisation and enlightenment that have made it possible.

As long as violence has not been eliminated, which I don't think it ever will be, there's always going to be enough there to fill the evening news and so if your impressions are driven by the events — and we know that it's a property of human cognition that our sense of risk is driven by memory, memory for recent and vivid and dramatic examples — unless you actually consult the statistics, you look at the denominator of the fraction, all of the people that aren't victims of violence, you'll come away with a misleading estimate of the risk.

* * *

Paul Ehrlich was such a star that the then leader of the New South Wales opposition told me that if he, Bob Carr, ever became premier, one of his first acts would be to introduce Professor Ehrlich in a public forum to talk about population and the environment. And he kept his promise.

Paul talks without notes, strolling around the stage, saying rude things about economists (Q: What do you say about a busload of economists driving over a cliff? A: A good start.). But like all our stars, Paul is not just a provocative orator; he is a leading researcher on butterflies and birds, doing extensive fieldwork every time he comes here, and is a leading member of the American Academy of Science, its top body for such scholarship. He is still treading the boards at nearly ninety.

You never hear any discussion, or virtually never, in the United States of the population problem. It's slowly starting to come back, but even then they say, 'Well, we're just going to add 3.5 billion more people to the planet by the end of the century.' What don't they mention there? Well, they might mention, for instance, that when I was born there were only 2 billion people on the whole planet, so 3.5 billion is a lot more than there was when I was born. We seemed to have enough people around to do things but more important they don't mention that that 3.5 billion may very well put more pressure on the ecological systems of the planet than the 3 billion or 4 billion that came before them. Why? Because every person you add disproportionably adds, on average, to the environmental impact: water has to be transported further; you have to win your metals from scarcer ores, you have to drill deeper for everything and so the disproportionate impact of this 3.5 billion people is going to be ginormous.

All those BILLIONS. The optimists hope that the global population will level out by 2100 at 9 billion. The pessimists suggest 11 billion. However it turns out each extra person, and we add about 90 million a year to the overall figure, increases the ecological pressure on the planet.

★ ★ ★

There is, of course, an abiding problem with Brits. Both Norman Swan and I were educated in UK. So were many of our stars. There is something about the written language that was encouraged when we were at school and university in the fifties, sixties and seventies. A well-fashioned sentence makes for a good radio talk. Our young scientists are catching up. So here are some more Brits, albeit with strong Oz connections.

Dr Paul Davies did research at Newcastle-upon-Tyne, Cambridge, Adelaide and now Arizona. He still lives in Sydney during parts of the year. His knowledge of astrophysics and those tiny particles at the other end of the scale is really AWESOME. I have never seen him flummoxed by any nuclear or stellar topic. And, despite his shyness of public intrusions, he is ever willing to come on *The Science Show* to explain some new conundrum.

> We don't know where life originated but my guess is that maybe deep under the ground, perhaps, where the internal heat of the Earth mixes with sea water, we know that there's primitive life down there. If you drill into the seabed and you look at the genes of these organisms, they seem to be like living fossils. For my money, that's a pretty good place to cook up life.
>
> I've never thought it was really very sensible to suppose that life happened just once. Now when you look at the history of the Earth for about the first 700 million years, asteroids and comets were bombarding it quite mercilessly — conditions rather like hell. And pretty much as soon as these conditions abated, life was well established. So it happened pretty quickly and that sort of suggests that it happens quite easily. We can then use those facts to make a numerical estimate of just how likely it is that life happened more than once. If these two forms of life then get in each other's way, it's a matter of simple mathematics. And it turns out that there's about a 95 per cent chance that life would have happened more than once. I think we should go and look for signs of a second genesis here on Earth.

The extraordinary thing now is that top oncologists in America have recruited Paul to help shed new light on cancer, the argument being that astrophysicists look at issues from the ground up, with no assumptions beyond strong evidence, and then develop theses to explain what may be happening. Well, Paul agreed, on the basis that many cancers were becoming too common, enlisted Dr Charley Lineweaver, another physicist from the ANU to collaborate, and they have been shaking the cancer foundations now for about two years. Their ideas, in brief, are that cancer cells are the leftovers from our ancient past. They differ from normal cells, which are willing to cooperate to form bodies and even to die voluntarily (apoptosis) to make way for fresh tissues made up of younger cells. Cancers, on the other hand, will suddenly be triggered to go their own way as individuals, damning the rest of the body, just like free-living microorganisms do. The way to treat them? Starvation: they are especially vulnerable to depletions in food and oxygen.

> The cost of cancer research is really quite staggering. The United States spends $5 billion a year of taxpayer money, and then there's charitable money on top of that. Just to put that into perspective, that's more than the US government spends on space exploration. And the total price tag over the duration of the war on cancer is about $100 billion. This is just the US. There will be comparable figures, say, for Europe and other parts of the world.
>
> I came into it cold. And the whole idea was to bring a fresh perspective to a very deep and complex set of problems. One reason to bring physicists into this field is that they can ask really stupid questions, seemingly without embarrassment. And sometimes those really stupid questions, like 'Explain to me how that works', and, 'Supposing you ran it backwards', and, 'Would it matter if you turned the temperature up a bit?', — to all these sorts of questions cancer biologists often say, 'Nobody has ever asked that', or, 'That's a very good question', or, mostly, 'Well, nobody knows'.
>
> And I always figure if you get a lot of clever people in a room, the sum total of what everybody doesn't know is what is really interesting because that's where progress is likely to be made. That is the opposite of brainstorming — you get people together and you say, 'Let's pool our knowledge.' I'm after pooling ignorance because it's in the gaps of ignorance that we are likely to make progress.

The best I can do is bring a lifetime of experience across several sciences and survey the scientific landscape and spot what I think are significant facts — almost all of which will be known to somebody in the cancer community — and connect the dots in perhaps a novel way. I'm trying to change the culture of thinking, the culture of research in cancer biology.

I think we have to position cancer in the great sweep of the story of life on Earth. So here they are, the questions: What is cancer? Why does it exist in the first place, and how does it fit into the great story of life?

The take-home message is one that is both good news and bad news. I don't think cancer is a disease to be cured. I think it's a condition or a phenomenon to be managed. I think we can manage it, but the lure of the cure I think has distracted a lot of research. People want a pill to make it go away. I don't think there will ever be a general-purpose pill to do that. Cancer is not that sort of thing; it's not like an infectious disease.

* * *

Roger Short, at the University of Melbourne, is a posh Brit who went to school with John le Carré and came here to study sex. He has become wonderfully mischievous: wearing T-shirts bearing rude messages to popes about condoms, recruiting prostitutes to test simple contraceptives that may kill HIV (lemon juice!) and giving any number of bright students a chance to flourish. He has done so many broadcasts on everything from elephants to penis length, it is hard to pick an example.

I had this love, going back to Rudyard Kipling, of elephants. And working in the vet school, I had heard from several people that every scientist ought to have a hobby project in addition to your main project. I thought, my hobby project is going to be elephants, so why don't I take a sabbatical year's leave, which I was entitled to, and spend half of it in New York, working next to the Rockefeller Cancer Research Institute and going into Rockefeller every day, and the other half I'd spend in a remote little village in Uganda studying elephants. I learned more in the six months in Uganda than I did in the six months in New York. Studying those elephants was just so amazing. It really blew my mind to be in the heart of Africa.

I took my crossbow with me and I developed a dart. I think I was the first person to dart and immobilise an elephant using this newly discovered morphine analogue, etorphine, or M99, which was developed in England. I discovered that you could knock out an elephant with just 5 milligrams, so you needed only half a millilitre of drug injected intramuscularly. So I went about darting elephants. I put the first radio collar on an elephant.

When I went back to Cambridge, I found that lots of people were interested in elephants, so I got quite a few students who wanted to do PhDs with me on elephants. One of these, my first I think, John Hanks, went out to Zambia, where there was an elephant-cropping scheme, and started collecting data. The Zambian government objected to what he was doing, although the government were doing the culling, and he was put in protective custody. I had to try to get him released, and I didn't know what to do. I heard that [conservationist] Sir Peter Scott was visiting Cambridge and I thought, 'Peter Scott knows Kenneth Kaunda [president of Zambia] very well. Kenneth could get my PhD student out.'

So I asked Peter Scott if he would come to my office one evening. He came at about six o'clock and he said, and this is a verbatim quote, 'You know, when we started the World Wildlife Fund, its objective was to save endangered species from extinction, and I'm now near the end of my career and we've failed completely. We haven't saved a single endangered species. And if we'd put all the money we collected into condoms, we might have done some good.'

And I remember thinking, God, what a thought! Good heavens! Of course, he's right! What am I doing here wasting my time in a vet school teaching horseshoeing to vet students? I really ought to be leading a research group in human reproduction and seeing if we can get contraception working and out there and available to everyone. It's human population growth that's the transcending problem of our times.

A later interview:

RW: [There's a] really fascinating contrast between ourselves and our cousins, the great apes. Most men — though Queensland may be an exception — don't have a bone in the penis.
Roger Short: There is great variety among primates as to whether or not there is a bone in the penis. We don't have one at all. The gorilla has a teeny-weeny little fleck of bone, the chimp has a wee bit more bone, and some of the macaque monkeys have quite large bones. And this has always been a bit of a mystery, but I think the mystery has at last been solved by a primatologist in Britain, Alan Dixson, who has just demonstrated rather nicely that those species that have a bone in the penis are the ones that maintain an erection for an extended period of time after ejaculation. For example, in the dog, with this long post-ejaculatory copulation, you've got a very big bone in the penis. This also seems to explain why some primates do and some don't have bones in their penises.
RW: Is there any information on sensitivity? I suppose there wouldn't be much work done on that with primates but what about on us?
RS: It's interesting you should ask that. There has recently been published a study by some intrepid physiologists who

have looked at the sensitivity of the human glans penis, that's the tip of the penis, and they've found, contrary to what everyone might imagine, that it's extremely insensitive. They say it's comparable to the sole of the foot and certainly one of the less sensitive areas of the human body.

RW: And that's a paradox. One would have thought the reverse.

RS: Yes. It's interesting because actually this has quite a lot to do with condom design and development, which is exercising many people now with a view to trying to do something about the threat of AIDS and improving the efficacy of condoms as a barrier. As you know, if you talk to people about condoms you always get the same old sick joke about how it's like having a shower with a raincoat on. Well, clearly if the penis is as insensitive as it seems to be, then it probably doesn't matter too much just how thick the condom actually is as far as transfer of sensation.

* * *

Richard Dawkins lives in Oxford but comes to Australia a lot. I first talked to him nearly forty years ago when *The Selfish Gene* was first published. His voice was boyish then and is deeper now; but his statements can also be more, some say, strident; I will say unambiguous instead. He has given us the courtesy in both written work and in speech of telling us what he really thinks, not some bowdlerised version to be polite. This is especially important as the forces against both evolution and secularism gain strength, perpetrating the most appalling crimes around the world. How dare they try to trash Darwinian ideas on natural selection, one of the most elegant and important intellectual contributions to the truth in history? It is criminal to try to ban these ideas from schools and to ban textbooks that contain evolution. The scandal is that there are excellent aspects of evidence to counter arguments mounted by the intelligent design mob, but they are shamelessly ignored, just as climate deniers ignore the multiple answers to their naive critiques. I am not surprised Richard gets cross.

He is often ambushed by the press as well, telling me, for example, of a charming journalist from the London *Sunday Times* (Camilla Long) who does a straightforward interview with him and then portrays him as a vexatious clown in her long article. I would be similarly furious after such treatment, multiplied scores of times. I understand from one of her colleagues at the paper that she and other journalists have been under instruction to provide knocking copy however pleasant the encounter.

Criminal behaviour under the badge of some religion is more obvious. Erect a flag, cut a throat; shout a pious slogan and capture hundreds of schoolgirls. Quote a 2000-year-old book and stop young people obtaining contraceptives, often resulting in ill health or even death. Richard has not ducked these issues. Do it too much and you may sound like an inflamed zealot. Such are the hazards. On a cool day, Richard Dawkins is genial and collegiate. He will often have happy, friendly conversations with friends who happen to be priests. There are lots of them in Oxford.

> **Richard Dawkins:** I felt that there are rather profound questions, like what is the meaning of life — rather pretentious-sounding things like that, why are we here, what is man — which could be given a really quite simple and straightforward answer that anybody could understand and I was intrigued and fascinated by the sorts of answers that modern biology is able to give and I wanted to share some of that fascination.
>
> I'm not setting out to be antagonistic, I'm setting out to be true. I do feel passionately about what's true, as any scientist should.
>
> I am a Darwinian, and it very often is put to me: What is the Darwinian survival value of religion? I think it is

necessary that anything that is as ubiquitous as religious belief is ... it's like sexual desire, that's equally ubiquitous ... there are some exceptions; not everybody has the same kind of sexual desire, but essentially it is universal and therefore it does demand a Darwinian explanation. But the main point I want to make about that is that you don't necessarily apply the Darwinian explanation to the object of your own interest, which is, in this case, religion. I think you can build up a kind of Darwinian story, not just for why arbitrary rubbish survives but why particular kinds of rubbish survive, because they actually take steps, rather like genes do, to survive by killing the opposition, in some cases by being positively attractive, like the idea that you survive your own death. It's easy to see why that might survive. So you can put together a collection of stories about why certain beliefs spread rather than other beliefs spread, and some of those will have these extremely negative consequences, like killing the purveyors of rival beliefs.

RW: So what about a moral code? If it doesn't come from God, where does it come from?

RD: If the rulebook is what you find in the Old Testament or the Koran, it's a horrible rulebook, which I think most of us probably don't obey nor would wish to, nor would expect other people to. It's true, of course, that you can find verses in the Bible and the Koran which do say the kinds of things that nowadays we would think of as good buried in a whole lot of things that we would say are bad. There must be some criterion for choosing which verses of the Bible we do accept, like the Sermon on the Mount, for example, and segregating them off from the ones that we no longer do, which are most of the Old Testament.

If you're going to accept that nowadays we don't believe most of the Old Testament anymore, then you've got to answer the question: by what criteria do we decide which bits we don't believe? Whatever that criterion is, it's available to all of us whether we have a holy book or not. You might as well bypass the middleman, bypass the holy book and go straight for that criterion, whatever it is.

* * *

What can one say about my final star — one of the oldest but never to be forgotten, David Attenborough? The last time we met he was 88. I said: 'Do you realise, David, we have one hundred years of science broadcasting between us?' Quick as a flash, he replied: 'Yes, that means you must have done twenty.' David is as sweet as New Zealand honey, but his mind is as fast as a CERN particle. You find that in the field when he needs to get work done amid any number of blockages. He sums up the situation and gives calm orders about making the most of what is happening, and gets on with it.

Efficiency, insight, scholarship (especially on birds of paradise from Papua New Guinea), but he still calls himself, as I do, just a journalist. We report on the evidence.

Nonetheless, accumulating evidence has forced David to speak out on what we face. After decades of public reticence, he has found evidence of the threats to our natural world to be overwhelming. What he used to confide in private, carefully, he now says boldly and unhesitatingly — like this comment about clear and obvious threats to our own coral reefs.

David Attenborough: I lower my voice because the doom, the thought that the world might lose the Great Barrier Reef, because you're going to lose a universe really of every form of water life that you can think of. It's the rising sea levels that are causing these initial problems.

The sea level is only one problem. The even bigger problem is the acidification of the water, the increasing levels of carbon in the atmosphere which is absorbed by the sea in the form of carbon dioxide, which turns into carbonic acid, which changes the acidity of the oceans, which are key to the survival of corals. And if that acidification goes on it's difficult to know how the corals, as we know them, can survive. The point, of course, is that changes of this magnitude are just at a moment in the geological record; human beings are only here for a moment in the geological record. And as for my generation, your generation and our grandchildren and so on, that hardly registers at all. And within that timescale, this disaster could happen. And if the reefs were to recover, that's not going to be for centuries.

RW: What would you like to see happen or change to make them safe?

DA: Well, the key element is the amount of carbon in the atmosphere that is affecting not only the reefs but everything.

I think it's very disturbing indeed that when you get a message from science that you don't like you say, well, the science is wrong.

That's not what science is about. Science is about looking unblinkingly at the facts, and when they don't match what you thought, then you change the way you think, and they produce a danger sign. The same discipline, the same mental discipline said, all right then, we must recognise the danger and do something about it. You don't say the science is wrong.

The great scientists are the ones who see something that destroys a theory, which they may have held for all their life, but who then say, 'Yes, I recognise the facts, there is a demonstration of the proof, and I will therefore abandon my previous theory.' That's what a great scientist does. But it is easier for a politician to say, 'I don't believe it and therefore I'm not going to spend money there or take any action.'

We just have to keep on declaring the truth, and that's not just restricted to science. It's a basic thing of life; it is a moral thing in life. That applies to science as it does for everything.

9 Failures
The pleasing proof of risk-taking

My worst failure on air as a broadcaster was in November 1974. It was not in *The Science Show* but in a long, live program called *Investigations*, meant to train me as a presenter to handle all that's thrown at you in a busy studio. I had been live only once before, ever.

This program was about the 'media game' and had a kind of agitprop assumption that the ABC was pure and magnificent and that many commercials were corrupt, devious and obsessed with money. The event was 'organised' by Matt Peacock, hero of asbestos reporting and still with the ABC as a senior reporter on *7.30*. Our guests included Kerry Packer; P.P. McGuinness; Adrian Deamer, first editor of *The Australian* newspaper; and Peter Manning, one day to head ABC News and Current Affairs. And about twenty others. We had links to other nations, phone-ins (the second time ever on ABC Radio), and a minister from the Whitlam government.

The 'plan' was to throw to a willing agitator who would take on Kerry Packer and other capitalist stooges and expose their duplicity and scheming. So I said 'Bruce,' (not his name) 'what is so wrong with the "media game"?' Bruce gulped, flung a self-righteous question to one of our moguls, who promptly deflated the whole edifice of outrage with a calm combination of good argument and telling fact. We never recovered. Each sally was swiped away by our captive captains of the airwaves and papers. We foundered for a couple of hours and then the final hour was abandoned. We sat in the studio and had a drink — a lorra drink!

One of problems was that nearly the entire cast was pole-axed with grog even before we began. Our ABC managers looked shell-shocked and the cast was softly spoken, almost inaudible with embarrassment — apart from Kerry Packer, who was in no hurry to leave and talked in a friendly and sincere way about how the show could have succeeded. I have never forgotten his calm generosity.

We survived. The General Manager of the ABC, Talbot Duckmanton, had penned a congratulatory note, presumably referring to the first broadcast of *Investigations* the week before, saying how good it was for a team to try something new.

A few consequences followed from that experience. The first is that I never again trusted anyone's assurances about a tricky broadcast unless I have checked out the essentials myself a dozen times. Secondly, it is silly to dub Kerry Packer or anyone else as a villain before you start. I have had good, friendly relations with any number of figures you might see as unlikely, from former PM John Howard who is always warm and direct, his delightful former science minister Peter McGauran and former speaker Bronwyn Bishop who has always been happy to talk about music and science, once central to her then portfolio as Minister for Ageing. If my relationship is cordial from the beginning I will not let my job interfere — the interview may be vigorous but there must never be invective. And, finally, I learned to avoid unfamiliar territory unless properly steeped in research.

So, from failures can come successes.

* * *

Our next failure came, sadly, from within ABC TV, with nudges from elsewhere.

We receive unending complaints and many a conspiracy theory. About twenty years ago a vet with ambition wrote to *Media Watch* saying that Dr Jonica Newby was a promoter of manufactured pet food and essentially a running dog of capitalism, doing talks for *The Science Show* whose aim was to sell PAL and Kit-e-Kat. Sometimes these claims are so barmy you assume no one would take them seriously.

Media Watch did. At the time it was run by two misanthropes who seemed to have little compunction in taking down anyone in sight, especially those appearing to be 'friends', that is, ABC favourites. And this was our turn. We pointed out that the vet making the accusations had written that tinned dog food could cause a type of AIDS and that the only correct menu for Fido should be Raw Meaty Bones (the name of his 'movement'). Why were they taking this laughable smear seriously?

Jonica was bewildered. Yes, her former employer was a pet advisory firm for whom Jonica acted as a link to the public, giving advice on looking after cats, dogs, budgies and the rest: how to stop dogs jumping up on your front and being aggressive, whether cats should be allowed outside and what to do with caged birds that needed company. Standard stuff on animal behaviour. Nothing about tinned food. The outfit *was* funded by Uncle Ben's, a pet food company, but Jonica had no interest in such products *at all*. *Media Watch* insisted she did — a plant in the ABC to foster capitalism, no doubt.

It's quite amusing, now and then, for someone in the ABC, especially in Radio National, to be accused of reactionary tendencies. The trouble is, it was utterly and completely wrong. Nonetheless, Jonica was attacked on *Media Watch* three times. As a result, I was told by one of the judges, she was dropped as a winner of a Walkley Award for her series *The Animal Attraction*. Secondly, her series, received so warmly by listeners, was banned from receiving a repeat run on *The Science Show*. This was explained frankly by a manager as being out of fear that *Media Watch* would strike again. Fear as editorial policy!

All this came about over a bonkers complaint any professionals in the field could have shot down in minutes. Jonica happened to be a member of the national board of the Veterinary Association at the time, with excellent stature in the profession. Since joining the ABC she has won the Eureka Prize for science journalism twice (it is the supreme award in the field in Australia) and she has also won the World Prize for TV journalism for her *Catalyst* special on measuring changes in our climate. She is, as I've already mentioned, my partner. *Media Watch* has long had a far more civilised team who may, indeed, seek out ABC snafus but do so without venom and with solid evidence — in the best traditions of science itself.

It is instructive to experience attacks from a bullying TV outfit. You realise quickly how others may feel, especially if the accusations are false and you can do nothing about them.

* * *

We were once caught out ourselves, quite rightly, when one of our speakers turned out to have plagiarised her little talk from *New Scientist*. Now our practice, with no time or staff to do forensic investigations on our contributors, means we rely instead on published papers, the presence of vetted speakers at top conferences and the imprimatur of universities who introduce us to both professors and PhDs. But when a student emails and says she wants to do a little talk you have to assume both goodwill and good ethics.

A couple of days after our student's short piece went to air we received a note from a listener pointing out the blatant purloining of many paragraphs from the magazine article. Our interlocutor sent the same note to *Media Watch* who duly contacted us. We explained we often accept outside material and sometimes have to assume it is in good faith. Nonetheless *Media Watch* went to air with our botch.

The student, now a graduate, was surprisingly untroubled. We all do it, she explained, suggesting that most of her student buddies swipe paragraphs from all over without blinking — for essays, articles or whatever else arises at university. We were sceptical that plagiarism at that level could be ubiquitous; too abundant, maybe, but not the rule, surely. Now, in 2015, we know better. She was also less bothered by the scandal on air than the worry about her very rich and quite famous father finding out. He had bought her a house for her twenty-first so his princess could be comfy, and she was horrified that he may be cross with her. We were stunned.

* * *

Our biggest failure, however, in 2016 is the problem of succession. Now we are down to as many staff as will fit in to a dwarf's dunny. Norman is part-time and visits for a couple of hours a week. Lynne Malcolm (*All In The Mind*), David Fisher, my producer for years on *The Science Show*, Amanda Smith (*The Body Sphere* for a few weeks a year), Ann Jones (*Off Track*) are just about it. Others hover nearby, but really, that is the Science Unit, from seventeen staff ten years ago to four and a half in 2016. And most of us make Methuselah look like a teenager.

The joys of yesteryear, when some of the best brains in the country surrounded you, are long gone. We had Alan Saunders (philosophy, food, films, history, music); Johnnie Merson (now a professor of nearly everything at the University of New South Wales — UNSW); Robin Hughes (she ran Film Australia and is now the pro-Chancellor of the Australian National University, no less); Peter Hunt (one of the best-informed ecologists in the nation); Kirsten Garrett (former editor of *Cleo* and after us, executive producer of *Background Briefing*) and many more names, yes, I could go on. But the point is you could think of an idea for an issue and find a superbly informed person to discuss it with only metres away. That was the underpinning of all our programs. That intellectual heft has largely gone from the ABC. You may have some marvellous minds in some parts of Aunty, but most of us are too busy to squeak, staring at our screens with headphones on, let alone debate the meaning of life.

Without a succession, that is, young staff costing a quarter of a managerial 'suit' (who inevitably requires an office and a PA — we have neither), we are on the dirt road to oblivion.

All of which raises the question about why a topic such as science and medicine, always top of listeners' and viewers' favourites, should suffer such neglect? And not just in the ABC, which still has more science journos than anyone else. *The Australian*'s experienced science journalist, Leigh Dayton, has left to do a PhD at Macquarie University. Deb Smith, *la suprema* from the *Sydney Morning Herald* has joined UNSW as a publicist, Jo Chandler has gone from *The Age*, as has Wilson da Silva, former editor of *Cosmos* magazine, now at the Department of Engineering, UNSW. So the pattern is reproduced around the country. It is like a particularly rugged episode of *Game of Thrones* when all your favourite characters are garrotted by the end, but without the preliminary sex to make it all worthwhile.

One answer is the concept of specialisation. I wrote a book about this in 1996, *Normal Service Won't Be Resumed*, in which I forecast the collapse of specialisation as a result of two parallel forces: cost-cutting, and that great Australian standby (pace *The Lucky Country*) and default position, being anti-intellectual.

Science and science reporting are both like soccer — you can't do it well with only one type of player. Your team needs goalies and strikers and solid guys in defence. It is also, if you're not into sporting analogies, not a good match with onanism — doing it solo is unsatisfactory. So, as I indicated above, a really good science-reporting outfit should have a medic (Norman Swan), someone with green credentials (Peter Hunt, Ali de Blas), a stargazer (David Ellyard), a policy person (Peter Pockley), a psychological specialist (Natasha Mitchell, Lynne Malcolm) and a techie/engineer. Only three of these are left: Norman, Lynne and me (a generalist). We get away with it because all three of us are immensely experienced. But it can't last.

Editors in newspapers and commercial broadcasters prefer having a big pool of non-specialists because you can anoint Fred or Frieda next Tuesday to cover anything you like, from police rounds to the cure for malaria. NPR (National Public Radio in the USA) has more of that kind, with some science specialists remaining who feed into current affairs programs as required. No editor in that situation has to contemplate units with people like me who stay for forty-three years or Norman, thirty-four years. And the editor keen on mobility will add that the internet now provides unlimited scope for someone keen on a topic; just to look it up, not wait around for weeks in case *The Science Show* or *The Health Report* gets around to it.

So what would a media scene without specialisation and small permanent teams look like? You already know. Just turn on a few non-ABC or SBS services and it's already there. Yes, the croc, dinosaur and sharks stories will appear, but there will be little of your unexpected reports of Betty the Crow, dog evolution, telomeres, brain centres for addiction and all the stuff you find as a reporter by just being there for something else and then getting the tip-off, or having a colleague race in to say 'Guess what?'

* * *

There is another failure — one of communication and interaction. It used to be different; now it is a rarity. News and current affairs staff may have not one (the bliss of twenty years ago) but four or five reports to do on this mad 24/7 media nightmare. There is little real need to keep bashing out the stuff — how much do members of the public care whether they hear bad news thirty minutes earlier? And do we really need those content-free updates on the princess's non-birth or the election non-result, or the 105th priest to be got for fiddling with kids? These days I find I need to turn off news and current affairs programs because it's all so bleak, or empty.

But we have plans following the fortieth year of *The Science Show*. We shall invite some of the many ABC presenters interested in science to do occasional reports and be stationed in our unit. It will be stimulating, not just for us, but very much for listeners.

And no, I am not reflecting an attitude from last century. Yes, we can email and do screen chats with everybody from everywhere, but being *with* someone is different.

Consider a new book called *The Village Effect* by Susan Pinker, the sister of renowned Harvard psychologist Steven Pinker. She writes, 'We're lonelier and unhappier than we were in the decades before the internet age.' Somehow I feel we know this. I am OK because I get up off my chair when I want to talk to an ABC colleague. I walk down and up stairs. They are often pleased to see me and say quickly (and rarely) if they are flat out, but otherwise we have a brief but pleasant talk. For most, however, it's a case of being chairbound, and the cost to the employer turns out to be large. Susan Pinker writes: 'In short evolutionary time we have changed from group-living primates, skilled at reading each other's every gesture and intention, to a solitary species, each one of us preoccupied with our own screen.' She could have written ' ... our own mirror.'

And now the ophthalmologists are telling parents to ensure their children put down their screens and go outside to look at distant objects. Otherwise their eyes will soon be stuck in close-reading mode and they will be myopic. Exercising your eyeballs so they don't seize up — there's a thought!

A final failure. Ten years ago we had a weekly environment program. It was called *Earthbeat* and was presented by Alexandra de Blas. Before that, we had several — one was presented by Peter Hunt (was it really called *Earthworm?*). Peter was a magnificent mind with a brilliant training in geology from Macquarie University. Legend has it (actually it is true) that

Jill Wran was listening to *The Science Show* one Saturday and heard Peter's superb coverage of the protests at Terania Creek in northern New South Wales against logging. She fetched her husband Neville, premier at the time, who became transfixed. As he listened and thought, like the focussed QC he once was, he realised how absurd it was to ruin old-growth native forest to make woodchips then toilet paper for the Japanese or rough planks for some distant house. Wran changed New South Wales' policy to protect forests. He told me later (on a TV program called *The Uncertainty Principle*) that this was an achievement of which he was proudest.

Peter won so many prizes for his forensic, fair and influential reports. If the old foresters told him how the industry really worked, or should, it was in his coverage. You could not accuse Peter of being partial beyond the facts. Sadly, Peter's heart was, from birth, back to front. One day, setting off for work at 5 a.m., he collapsed and died. So young!

His was the type of program we still need, but there are no funds. So all the green stuff has to be in *The Science Show*. It is a great pity. There is no way I have the time (nor the ability?) to do what Peter did beyond a quick interview giving the topic a mention. We may do a bit more now and then but a separate, clearly independent and non-ideological program on our monumental environmental issues is sorely missing. Gregg Borschmann does well on *Breakfast* (Radio National) and *Background Briefing*, but we need more. Don't hold your breath.

10 Successes

How can you tell?

How hard it is to predict what works. I do something on Saturday, even now, wait until Monday and ... nothing. It is as if the show never existed. What is wrong? Have I lost any knack I once had? Another day passes. A euphoric note comes from Switzerland, then from New York and Manchester. I postpone suicide. Finally, a bunch of Australians send congratulations. I realise that not everyone is on a hair trigger as the show finally manages to make a mark about its possible worth. You just have to be patient.

But sometimes you know you've hit gold, however unlikely the hole in the turf.

In 1979 a round tin arrived from Perth. Its label said Richard St Barbe Baker, recorded by Barrie Oldfield. Barbe Baker, it said, was nearly ninety, had worked at the turn on the twentieth century with Baden Powell, founder of the Boy Scouts. Just what we need now, I thought (it was a rough day) — *into the bin*.

I then paused and retrieved it. My being thirty-five was no reason to jettison a fellow three times my age (these days there would be no hesitation from the demographic fascists). I opened the 7-inch battered tin and put the pink recording tape on my machine. There was Barbe Baker, out in the bush with the ever-so-polite interviewer Mr Oldfield, talking about planting trees in the desert. He recited a prayer as he re-enacted the ceremony.

As I heard the words I also, somehow, heard music at the back of my mind: Vaughan Williams, *Fantasia on a Theme by Thomas Tallis* — the two came together like a charm. I listened to the end — it was half an hour or twice the usual length for an outside contribution. This was looking beyond extreme!

I did some edits, found the music, and put the combination to air the following week. What a response! It was simply incredible. But why?

Well, first, Barbe Baker, even before editing, had that measured, enunciated British elocution that was a joy to listen to. Second, he was talking about marshalling the help of ordinary people like children and grannies to plant trees where they are not supposed to be able to grow in a thousand years, using mulch, stones, even oil residues from the wells nearby — slimy black slurry that's otherwise vile leftovers to be buried but now was being enlisted to create life. Third, his schemes were clearly working and, despite the evangelism, it was good news indeed. But above all it was *emotional*, moving. And then I added the Vaughan Williams and many a hard man listening turned to jelly. (The same piece was used by Peter Weir in the film *Master and Commander* when a seaman is lost [purposely] to the ocean and struggles vainly to survive.) As for the women — the organisation was called Men of the Trees, but only because they are so ancient. Everyone's welcome.

That day of the broadcast, in Melbourne, a fifteen-year-old boy was listening. He was totally enraptured. Scott Poynton, on the spot, changed his career plans and decided to become a forester. Thirty-five years later Scott wrote to me, telling me the story. He was emailing from Switzerland. He told me that his organisation is called The Forest Trust. One of the reasons he was writing at that time in 2014 was that he and his trust had just persuaded the main oil palm company in Indonesia to agree to stop their destruction of native forest and the ruining of the orang-utans' habitat. Could this be true? Enlisting the powers of Nestlé as customers for oils for leverage,

Me with Welsh forebears in the Rhondda Valley some 30 years before *The Science Show*: grandma Miriam and grandpa Johnnie Williams.

Posing for an ABC Radio 2 publicity shot in 1982.

On location in the early days – New Zealand 1976.

Liz Parer-Cook who with husband David did so much pioneering wildlife television production for the ABC Natural History Unit that is now sadly no more.

With Robin Hughes, who helped *The Science Show* into the world back in 1975, and the founder of the ABC Science Unit, Dr Peter Pockley.

Professor Peter Mason was our first real star and a natural broadcaster. He continued broadcasting with me even after his diagnosis of brain cancer.

With the Space Shuttle 'Full Fuselage Trainer', Houston Texas.

The irreplaceable Sharon Carleton on a visit to HRH Prince Charles in 1993 to discuss butterflies and sustainable sewage disposal.

Above: With frequent contributor Rod Quantock and frequent listener Ron Barassi at the launch of *Science Show II* in 1986.

Left: With Polly Rickard, *Science Show* producer before David Fisher.

Dr Oliver Sacks was a surprise visitor to our twentieth anniversary bash in 1995.

Introducing physicist Paul Davies at a Radio National 'Science Lunch' in 1993. Paul discussed theories concerning the origin of the universe and the possible existence of a creator.

Just a few of those who have kept *The Science Show* show on the road for forty years. Clockwise from top left: Johnnie Merson presenter of many series; Peter Hunt who changed history with his *Science Show* broadcasts; David Fisher (producer since 2004) on location in Germany; Tim Bowden, who presented many a brilliant piece and produced famous series about war and Antarctica for Radio National; Jonica Newby, expert in animal behaviour (and my partner).

On location, Machu Picchu, Peru, in 2014. *(Photo: Tom Williams)*

On the shores of the Galapagos Islands – where Charles Darwin's theory of evolution began to take form. *(Photo: Taro Regan Williams)*

Scott then focussed on the main culprits: 'Nestlé led Poynton to Indonesian/Singaporean palm oil supplier Golden Agri-Resources, which signed up with TFT in 2011, then to its sister company, Asia Pulp and Paper. APP had become a poster-child for everything that business could do wrong environmentally.' So wrote Michael Bachelard in the *Sydney Morning Herald*'s *Good Weekend* magazine in March 2014, telling the story of *The Science Show* broadcast and what happened decades later.

As Poynton said of Barbe Baker: 'He told these terrific stories, this old guy with this fantastic British voice, and he used poetry, he used science, art. He came at people from fourteen directions; he took me on this journey.' Poynton sent me a sculpted tree, no taller than a tin of beans, exquisitely using its trunk as a lovers' embrace and its branches spreading in coppery elegance. It was a thrilling present and an unbelievable reminder that you never quite know the consequences of your broadcasting. And in 2015, thirty six years later, Scott Poynton gave special thanks in the acknowledgements of his book on forests, *Beyond Certification*, to the influence of that broadcast.

I'll count that as success, a reward for taking the trouble to listen to someone's (Barrie Oldfield's) contribution. Another success involved Prince Charles, the GG and fleas.

I was keen for our supremely talented and resourceful freelancer (and dear friend) Sharon Carleton to do a portrait of another toff, one I'd seen in Britain several times wearing her flowing robes in autumn colours (she insisted on two sets of clothes, one for summer and one for winter, that took no more than 45 seconds to put on), all the while telling the world about her fleas: Miriam Rothschild. Rothschild had little formal education in science. She loved her fleas because they were so lively and had unexpected relationships. She lectured around the world as a respected expert and the rest of the time floated happily like many a Brit eccentric.

But gradually her eyes got weaker and she could no longer see her charges. So she went up a scale and turned to butterflies.

Prince Charles heard of her work and decided to seek her advice on what plants to put into the gardens at Highgrove, one of his estates, to attract the glorious creatures. Rothschild agreed. The experiment worked and the Prince of Wales was well pleased.

So, as part of our *Science Show* portrait I thought it would be fun to interview Charles. Two problems arose immediately: no funds to fly to the UK and no access to a prince besieged by a nasty press exposing his love life. So I called the Governor General, my friend Bill Hayden, who used to phone regularly to discuss which science books he might read. He contacted the Palace who said ... OMG ... 'Yes.' Now we were on the spot. What about the airfare? Here Sharon stepped in and sweetly announced the availability of husband Richard's millions of frequent-flyer points accumulated with *60 Minutes* on Channel Nine. Off she went. This how she remembers the encounter today:

> We'd been given detailed instructions from an equerry before being ushered into the presence of Prince Charles at Kensington Palace in London. 'You may bow or curtsey, say Your Royal Highness first and then Sir afterwards.' But there was no time for any formalities. HRH strode up to me, hand outstretched: 'Good morning, Mrs Carleton, you didn't come all the way from Australia for this did you?' I laughed, and told him I couldn't think of a better reason, and did he mind if we got the hard bit over first, the photographs? *Woman's Day* was on hand for the occasion. He looked at me and said with a grin: 'You know being photographed with me could spoil your reputation?' I rather hoped it would.
>
> The equerry had also given me strict instructions that I wasn't to stray from the precise subject of Miriam Rothschild and her wondrous planting of wild flowers at HRH's country estate, Highgrove, in Gloucestershire. I didn't, but he did.
>
> Knowing exactly what he was doing, Prince Charles gently led our discussion into areas of sustainable waste disposal and dragonfly ponds, restoring the English meadows

and wetlands and then back to Miriam Rothschild. We made two programs featuring him.

After I left and was walking across the Palace forecourt, the equerry called after me. 'Seeing as you're doing a program featuring the flora at Highgrove, Prince Charles wondered if you would care to visit there yourself?'

'Yes please.'

Being on the spot! How success stems from being there. The program was a delight, of course. And here is a sample of what Prince Charles had to say.

> We have managed, since the last war, to destroy vast, vast areas of this country in terms of its flora and fauna, all the herb-rich meadows, the chalk downland and the wetlands. We have actually reduced these areas to very, very small patches here and there.
>
> I feel very strongly about trying to make my own tiny little contribution to try and find ways of restoring some of these habitats. The difficulty, of course, is to try and regenerate such places because once you actually plant them up and have destroyed an ancient system of management — because most of these areas actually produced the interesting wildflowers because of the type of management, grazing or cutting for hay — once you stop all that and you introduce a different system altogether, it takes a long, long time [to regenerate]. It's like once you start destroying or cutting down ancient woodlands you can't just recreate an ancient woodland — it takes hundreds of years.
>
> **Sharon Carleton:** Sir, I believe Miriam Rothschild also advised you on your dragonfly reserve, which I think you call by another name?
>
> **HRH:** The splendid German character who designed it for us called it a 'sewage garden'. He sent the most wonderful letter saying, 'Now you are the proud owner of a sewage garden.'

Basically, the idea was to try and do something about a disintegrating septic tank which we had. It was beginning to rot and a lot of the effluent was going down from the farm and from this tank, down a ditch into the river, which I thought was not the best thing, to say the least. We had to do something about it and we looked at a whole series of options. I didn't just want to have another normal, standard, conventional septic-tank system. Nor did I want a biological system, which sounded quite a good way of doing it but would then have required a lorry to come along about every six months and suck out this stuff and take it away. Now where the hell are they going to put it? It seemed to me by far and away the best thing was to try and create a cycle where you would contain everything and not waste the waste and ultimately turn it into something worthwhile like compost.

We finally discovered this German gentleman, Uwe Burka, who designs reed-bed systems to deal with effluent. It works on an ancient, very ancient, system and principle. In the old days it used to be based on a gravity system to get the effluent up into the reed beds and then flow down through the reeds and a willow bed and into a pond. That is if you've got height, if you've got a gradient. Of course, we haven't, it's all flat with us and I've had to put a pump in but basically it is, I think, a more self-contained system. It's more environmentally friendly in the long run and it can look rather nice as well and have this end product that you can put on the garden.

The trouble is the capital costs are quite high, much higher than to put in a conventional system. But I believe in the end it's a better way. The water quality ultimately which flows out of the ditch at the end, is now very clean.

Miriam [Rothschild] persuaded me. She said it would be a very good idea to establish dragonflies in this pond and it would be great fun to see if we could try to attract all fifteen species of dragonfly which exist in this country so I said, marvellous, what a good idea.

My own encounters with royalty haven't always been so smooth. In 1987 I did the Australian five minutes in a monumental 'as live' tour of fifteen nations called *World Safari*, hosted by David Attenborough and Julian Pettifer in London. It went well, though I noted that Prince Philip's contribution, from China on cranes, was entirely on tape. A few weeks later I saw him in the audience in Perth as I was MC for a major IUCN (International Union for Conservation of Nature) conference. On suddenly being asked to fill for a few moments from the stage as the backstage folk fixed a pulley, I found myself telling the story of the astonishing broadcast involving Jacques Cousteau, Rajiv Gandhi, Nikolay Drozdov, David Suzuki and the Duke of Edinburgh, and telling him 'But you, of course, Prince Philip, cheated a bit by introducing those birds on tape, not live.'

He was livid. 'Why am I always the mug?' he erupted when I saw him afterwards. It looked like a week's diplomacy with him and the federal government had been torpedoed. In fact, as I learned, Philip's moods are fleeting. This was confirmed during one of our real successes involving a Devonian fish that had invented sexual intercourse. Or, to put it another way, a fish in which Professor John Long, now of Flinders University in South Australia, had found a uterus. The tell tale signs were unmistakable. Such organs require internal fertilisation — hence sex.

We had set up, through the Australian Science Media Centre, a live press conference from Adelaide to the Royal Institution in London, with Long and the other authors of the *Nature* paper announcing the discovery with us in Australia, and at the Royal Institution were the Queen, two dukes, Baroness Susan Greenfield, the South Australian premier and any number of other celebs, not least Sir David Attenborough, after whom John Long had named the fish.

We were not allowed to involve the Queen in the Q&A with the scientists as this was all about sex but when the dukes arrived anything was possible. So there was a bit of byplay with Philip at the end of which I asked him, via satellite, 'do you have any further questions, Prince Philip, about our unique Devonian fish?' 'No' he replied 'but do give it my best wishes.' I responded 'I'm sorry, Sir, I'm afraid you're 365 million years too late. The creature is an ex-fish.' The dukes, Edinburgh and Kent, departed chortling. I broadcast the entire encounter in *The Science Show*, showcasing a marvellous Australian achievement plus a bit of royal jollity to give some relief from the agony of The News.

Since then, in yet another paper in *Nature*, John Long has revealed that the first fish had to do it *sideways* — like two salmon on a dish they had to juxtapose their flanks as that is where the required apparatus of reproduction was placed. No missionary position or doggy style for those ancient ichthy-lovers!

* * *

Another of our successes has been finding ways around our increasing impoverishment. I mentioned before that David Fisher and I do *The Science Show* fifty-two weeks per year. Our colleagues at the BBC or CBC would need at least five bodies — and a few weeks' break at the end of the year. Without reporters whom you could commission to follow up stories in a systematic way, we have freelancers who suggest what they want to do. This is an OK if somewhat more random, meaning that essential stories are often done by us in a rather cursory way (as I have explained before) but the unexpected, like Richard St Barbe Baker, can have its unique benefits.

The exception, in a way, has been Sharon Carleton's marvellous sequence of portraits. Just look at this amazing list: Sir Jack Eccles (winner of the Nobel Prize for his work on synapses in nerves), Elizabeth Gould, Lawrence Bragg, Charles Babbage,

Barbara McClintock, Miriam Rothschild, Prince Charles, David Allen, Edward Jenner, Howard Florey, Mac Burnet, Pansy Wright, Ada Lovelace, William Perkin (and the colour mauve), Rachel Carson, Barry Marshall and Robin Warren, Sigmund Freud, Alan Turing, Hedy Lamarr, James Clerk Maxwell and Beatrix Potter.

Just a couple of thoughts on those names (easily looked up) — they are all Nobel laureates or film stars, inventors and so on. Ada Lovelace, as I mentioned in the chapter on women, assisted Babbage in the invention of computers and was the first programmer. David Allen was one of our great astronomers — I was with him when he died, still watching the sky via TV as a comet crashed into Jupiter. Gould was a superb artist somewhat eclipsed by her famous husband, John. Barbara McClintock's work on jumping genes was scoffed at — until she was proved correct and got the Nobel at over 80.

Sharon's journalism, writing and delicious voice has truly been one of our great successes. May she continue forever!

* * *

The ABC is a bizarre place. It has its history embodied by great broadcasters who have left the organisation, such as Tim Bowden, Chris Masters, Jonathan Holmes, Caroline Jones — you know the names. It has its regime of masters from the board and directors of 'platforms' (as they call them these days!); the scores of craftspeople from sparks to despatch without whom we could not operate; and people like me, who just broadcast. Sometimes, out of that uncoordinated mix, comes policy that gets sent around as PowerPoint performances that a manager will present in ABC buildings around Australia, whereas some simply emerges.

Our complement of science spruikers simply emerged. To understand this you must recognise that there are three different kinds of science communicator, all of which the ABC has in abundance (almost by accident). I would prefer to say it is the natural product of having hired a good range of talented young men and women way back.

The first kind of science person is the teacher. This comes from a tradition going back to Julius Sumner Miller who came here from America and demanded 'Why is it so?' in between promoting chocolate and its 'glass and a half of milk'. Sumner Miller was utterly didactic, and, as he confessed in his memoirs, something of a classroom tyrant. There was one correct answer to one of his shouted questions and, if you got it wrong, you were in strife. Our present generation of 'teachers' who similarly offer information in 'answer to questions' now use humour and fun to do their work, with marvellous success. They are Dr Karl, Bernie Hobbs, Adam Spencer (when in the building) and, in times past, Dr Paul Willis, ex of *Catalyst* and many *Science Show*s.

The second kind of communicator is the artist: the feature film-maker, the radio producer who mixes music and poetry and words from all over to make a work of art that has a huge emotional component as well as straight information. Sharon Carleton did this with Beatrix Potter. I tried to do it with features from Galapagos and Machu Picchu. David Attenborough does it with glorious films too numerous to count — and David Parer and Liz Parer-Cook have done it on ABC TV with mesmerising documentaries such as *Parrots of Australia* and *Wolves of the Sea*.

Then, thirdly, there are the journalists, who go out there and report what's new, what's important. They include Norman Swan, Lynne Malcolm, who often does *All In The Mind* shows of considerable emotional impact, as does Amanda Smith with *The Body Sphere*.

So who are the next generation of 'teachers', scientific artists and journos? Any names come to mind? I can think of many, but they are not in the ABC. Yet, as we stand, we have a good balance of these skills, to which we can add the on-line science communicators in the ABC, led by Ian Allen and including Stuart Gary's unique *Star Stuff*, presenting regular updates from the cosmos. (Both have now left the ABC.)

11 In the Beginning

How did the *Science Show* start?

I have mentioned previously that my being in the ABC, and even the endurance of *The Science Show*, are flukes. I was very lucky to get myself into the Science Unit when I did; lucky to be nurtured in the crafts of radio and television; and lucky to have a program supported for four decades.

There are lessons involved daily, of course — later in this book I propose that you get a good mix of possibilities and the coverage of what the public expects if you have taken the trouble to hire fresh talent that can emerge to fill the roles that new, often unexpected, demands will make.

In the 1960s I came to Australia for ten pounds on a ship called the *Castel Felice* (Castle of Happiness) with the worst food served outside the cheaper HM's prisons and accommodation best described as floating hovels — but what do you expect for twenty bucks? (It is one of those delicious coincidences that the science writer and journalist Victoria Laurie, now living in Perth, was on the same ship, landing on the same day as I did back in 1964). In the cabin next door was Delphine Matthew who had worked in BBC TV Music as a producer. On landing in Sydney, after a brief jaunt to the Snowy Mountains to earn my fortune (I lasted six weeks!), she introduced me to various ABC people she had met through the amateur drama company and other connections. This led to meeting my first wife,

Pamela, whose job was arranging top-line concerts in Sydney Town Hall, with names like Daniel Barenboim, Victoria de Los Angeles, Ravi Shankar and Tamás Vásáry. This was 1965, before the Opera House had opened.

In 1966, sticking to my plan to see the world, which was my original intention on setting out for Australia (never intending to stay), we hitchhiked to London. We set off from North Sydney and arrived at Piccadilly Circus nearly five months later. Once there, I enrolled at the University of London to do a degree in biology while Pamela joined Lew Grade's ATV as a casting director (the first series of *The Prisoner* with Patrick McGoohan) and then the BBC. While I was studying I did lots of TV; Pamela could get me plenty of willing agents handling small parts. So nearly every week, certainly during the long university holidays, I would be in BBC, ATV or other studios and on location in remote forests being the animated prop in series such as *Monty Python, The Goodies, Z Cars, Dr Who, Dixon Of Dock Green*, and one-offs such as Dennis Potter plays (*Son Of Man*), innumerable costume dramas and anything else that turned up in that 'Golden Age' of telly.

However, much as people may gasp at my being in *Python* and a couple of other now famous shows, it was no big deal then, in the beginning. These were odd, almost experimental productions, with *Python* not even having a clear program brief on being launched, just a collection of cerebral Oxbridge chaps with mad ideas. At first they were greeted with little more than an amused shrug. The cult status came much later. But for me, mucking about in so many varied shows was very instructive. I was appalled by the tyranny of the regulations imposed by both unions and the organisations: breaks after rigid times elapsed, food caterers on location however difficult to provide (at vast expense — could we not bring sandwiches?), preposterous overtime provisions, with shows risking closure if adherence wasn't possible. Some cast members, often old-timers, would be walking legalistic time bombs behind their whiskers and reminiscences about filming in ancient times, when *they* were nearly stars and about *theatre* when it was 'done properly', way back when.

Amid all the box-ticking, films and TV got made, somehow. I was always welcome because I learned to be on time, was briefed, however tiny my part, and unconcerned about rule number 387, para 6. My priority was the production, doing it well and doing it on time. Little did I know then how much this concern for professionalism was going to be the basis for survival in ABC Radio. I did not perceive myself as a natural-born talent. So, having landed in the ABC, maybe I could get away with staying there by being hardworking, reliable and not being bothered by daft rules.

One of the first rules we in the Science Unit were working around was editing. We were supposed to use technicians, however trivial the 'um' or 'er' needing to be removed. Then there was the requirement for those on air to be different from those producing the programs. We cut edited on our own machines in the office, out of sight, naturally, and used our own voices. This halved the cost of production (all quite illegal, at the time, of course).

My honest toil seemed to work as no one, despite my semi-freelance engagement from early 1972, suggested I go away. In fact I enjoyed the mentoring I was clearly getting at that time. Dr Peter Pockley, the Science Unit director, was insistent on high standards in the depiction of science, in the use of new technologies and in our being there for the *whole* of the ABC, not just our section of highbrow wireless. I learned a lot very quickly, particularly on how to concentrate on my skills, which seemed to be growing (editing was compulsive) and waiting for other skills to emerge in their own time.

I was appointed to a staff job on 2 December 1972, the day Whitlam was elected, though David Ellyard, with no broadcasting experience, got the senior position because (as it was revealed later), it '...was impossible to get Williams to dress appropriately', and I appeared to have '...scant regard for rules and regulations and often seemed to be taking the piss' — not exactly what they said in my file as revealed years later, but close.

Then Dr Peter Pockley left, or, in a way, was fired because of his own rule-breaking. He had spoken out about TV science, which was supposed to be under his command but which was never allowed near his grasp. TV managers were bothered by his memo prolixity and tendency to over-prepare for large projects, driving himself and others around several bends. Dr John Carmody, the physiologist and music critic, asked Peter a Dorothy Dixer about TV in public and, being at an open science congress hosted by the University of New South Wales, Peter thought he could not possibly duck the question. So he didn't. He revealed all.

He was carpeted by Sir Talbot Duckmanton the following day. Why is ABC science always so contentious? From that day to the present — funds, staff, closure, public regard, internal disdain — the saga continues.

With Peter Pockley departed, a new director stepped up from the EP (executive producer) position. Dr John Challis was trained as a philosopher — at the Vatican! He had entered the ABC in Education, then a prime entry point, and been grabbed by Pockley after a few TV and radio appearances in which John had displayed his deep intelligence, brevity and range. (By the way, Dr Challis's notoriety just recently in the newspapers in June 2015 came from speaking at a same-sex-marriage rally in Sydney, in tandem with Tanya Plibersek. He told a story about hiring a female hooker when young to test any heterosexuality he may have had remaining and then doing a runner because he fancied the girl not at all, so proving that gayness was naturally occurring, supposedly.) John then set about trying to work out what needed to be done in the Science Unit.

We used to have, up to the early 1970s, a topical science program called *The World Tomorrow* presented by the rough diamond Kiwi, Michael Daley. Michael had no degree; Pockley was a DPhil from Balliol, Oxford. Daley got pissed in the afternoon; Pockley was prudent and abstemious. Daley was an old-fashioned journo; Pockley was a toff from Geelong Grammar (at least in Daley's eyes). In other words, they hated each other. This was made much worse when Daley crossed the Harbour to Gore Hill to take charge of TV Science.

With the absence now of a topical science program, John Challis decided, with the help of Robin Hughes who was an honorary member of the Unit, to create a new one. Robin Hughes, as I have already written, is one of the most formidable minds you will ever have the good fortune to meet, and one of the most articulate. Her later experience running Mike Willesee's documentary team, then Film Australia and now effectively deputy to Gareth Evans as Chancellor of the Australian National University, tells all. We have always got on famously and I think she had an extra sympathy for my attitudes to conventional behaviour. She was also surrounded by media expertise. Her sister-in-law is Margaret Throsby, the great ABC Radio (FM) broadcaster, and her husband, Professor David Throsby, is one of the world experts on funding for the arts.

They got together to talk about possibilities. I (and others) had always fancied mid-day on Saturday as a time slot. Records of older music were then played there, plus an inordinate recitation of river heights. Robin suggested me as a presenter/producer. I discovered this only in May 2015 when I asked John Challis about *The Science Show*'s early history.

John then went to see Arthur Wyndham, then head of what was to become RN. So we had that triumvirate — John, Robin and Arthur (father of the *Sydney Morning Herald* editor of books, Susan Wyndham) — there at the moment of conception. They agreed that I be brought along in preparation for the birth by doing a series of long (three-hour) experimental programs in late 1974 called *Investigations*, the second of which was that spectacular failure *The Media Game* when Kerry Packer came to play (see pages 138–9 in the previous chapter).

Somehow I survived. When 1975 arrived, I gradually built up more current-affairs-style reports for our various other shows — *New Society* (presented by Julie Rigg), *Innovations* (a rather beige set of promos for gadgetry) and the forerunner of Phillip Adams's *Late Night Live*, *Lateline* on ABC radio, presented by Malcolm Long and a vehicle for some of those longer pieces questioning the role of science as elitist and narrow (see Chapter One).

And so came August 1975. We decided I should launch *The Science Show* No.1 from a huge conference in Vancouver, replete with superstars. I would go alone, as I had all the skills — writing, interviewing, editing, researching — that anyone would need and I could drop in at Stanford and Berkeley on the way. The occasion was so unusual in ABC Radio that I was given a party before I left on the roof of 171 William Street where many radio-talks departments were based (one of the first reporters I met there in 1972 was an AM staffer named Bob Carr) and I was given essential advice on expenses (claim *everything*!) and protocols in America. I travelled first class because all public servants did so then. You'd be up front next to generals, majors and tax inspectors. Minister Clyde Cameron changed all that a few months later when he reduced us to peasant class unless you are in ABC News/Current Affairs, when you go business.

I was not nervous of this challenge — on my own, in countries I had never before visited (USA & Canada) — but I was apprehensive about clashing with the EP of the CBC (Canadian Broadcasting Corp.) team, given our utterly different approaches to the craft: he with his team of six to eight; me with my team of none. They were doing their own shows and I was simply allocated a desk near them as we both covered this major event. Should he really be the one feeling edgy, I wondered, with his bounty of resources? But, no, he was on home territory, had massive experience, and I was a 31-year-old chancer, doing it for the first time.

As it turned out, we got on famously. We swapped tips, carried messages for each other, and proved once more that science reporting is charmed and friendly. On the day I was recording my script and sending the complete *Science Show* No.1 from the CBC studios, then at the Hotel Vancouver, everybody was immensely helpful. Given our typical lack of publicity back in Australia for the new program, I bought a small ad in *The Australian* newspaper announcing our arrival. I thought John Challis may have given me a few dollars towards the outlay, but he can't recall doing so.

Two months later, an academic from the University of British Columbia trod the same path to the same studio at Hotel Vancouver. David Suzuki, then a professor of genetics with some experience of radio and TV, carried his script for the first edition of *Quirks & Quarks*, a science program for radio of exactly the same length as *The Science Show*, to be broadcast at the same time (midday on Saturday) and, initially, with the same repeat slot. They had no inkling of my existence, nor I of theirs. It would take another ten years for us to meet, and then with far-reaching consequences.

Q&Q is still going strong. We remain the best of friends.

* * *

I saw David Suzuki for the first time in a bar in Toronto. He still took a drink back then. He was warm, informal and committed to broadcasting science that was bold, even uncompromising, on racism and environmental issues and the future. In ten years he had become, if not the most famous Canadian, then close (Joni Mitchell, Pamela Anderson and K.D. Lang were still out there). I invited him to visit Australia to do some lectures. He said immediately that the scientific establishment was annoyed by the line he took and I'd be risking opprobrium if we went ahead. I told him my middle name was 'risk' and he promptly made plans to come south.

He was then unknown in Australia. I wrote to the publishers Allen and Unwin (my friend, Patrick Gallagher) that I had a new colleague with a Japanese name who was a geneticist and did bits on Canadian radio (was there such a thing as Canadian radio?) and he'd written an autobiography and you, Patrick, should publish it! This unlikely nudge received an incredulous reply to which I answered, 'He's coming to Australia soon — trust me.' A&U signed up Suzuki. Several million book sales later, Patrick had this sweet, satisfied look on his face (apart from Suzuki, they also had *Harry Potter*). Sometimes it is rather nice to be taken seriously.

It wasn't so sweet with Barry Jones. Our science minister had set up the Commission For The Future to explore possibilities for a new century (this was the 1980s), not to bolster some Canadian iconoclast who minimised scientific ingenuity and maximised environmental hazard. And we had brought Suzuki out under the banner of the Commission. But Dave certainly got people talking. I broadcast his talks on *The Science Show* and he quickly became a superstar.

> I believe that the most pernicious myth we face in society today is one that is held as a truth by every politician, by every economist, by every business person in Western society and that is that in order to sustain the quality of life in our society we have to have steady, continuous growth. As you know, steady continuous growth, whether [it's] 0.5 per cent a year or 7 per cent a year over time, is called exponential growth. We believe that in order to maintain progress and the quality of our lives we must have sustained exponential growth.
>
> It flies in the face of any knowledge of what the world around us is like. Nothing in the universe continues to expand exponentially, indefinitely, nothing.
>
> As a biologist I look at the perspective of the entire history of our species, which is 400,000 to 600,000 years. If you plot on a curve 400,000 to 600,000 years on the x-axis

and you plot the distance we could travel in a day, the speed we could travel, the distance we could communicate, the speed we could communicate, the amount of air, water and food that we use, the amount of pollution that we produce, our numbers, for 99.9 per cent of our history that curve is flat. In fact, the normal condition for human beings is total equilibrium, is non-change. It's only in the last few centuries that you begin to see the curve go up and then in this twentieth century of course that curve is shooting straight off the page.

We only look at the recent past, the exponential growth, and we say, 'That's what's been happening, look at all the enormous material wealth that we have.' To maintain that we have to keep steady exponential growth.

If you look at the consequences of exponential growth you recognise very clearly that you can't sustain that. When the average North American consumes as much as 150 Indians (consume) per person, and we feel we have to keep growing even further. How can the planet sustain that kind of consequence?

It took all of human history to reach the first one billion human beings on the planet. In the last century then, in only 150 years, we doubled twice to reach four billion and now we're going to double again to reach eight billion in another thirty years. That's the consequence of exponential growth.

Our history as a species has been that we have sensed the environment is limitless and endlessly self-renewing. And for 99.9 per cent of our history it was. Our numbers were small, our technology simple and so nature could absorb whatever we did. We got used to leaving our leftovers out in the wild because nature would eventually reclaim them. The problem today is that our technology is so powerful that we assault the environment in a way that can no longer sustain the recovery. Nature now falls before us rather than renewing itself.

> We assume that nature can absorb everything that we do and yet everywhere around the screams from nature are warning us that it cannot go on.

The Suzuki Show in Oz has continued ever since. Now I find his unwavering encyclical predictable and closed off from much of the real possibilities of science and innovation unfettered. He is myopic on genetically modified crops (and this from a professor of genetics) and his condemnation of private enterprise sounds more like rhetoric than analysis. But I still love the guy. On seeing him in Vancouver to record a celebration interview to mark the fortieth anniversary of *Quirks & Quarks* he was helpful, generous and as delightful to be with as during that first meeting thirty years ago. I did get the feeling that Dave was tired of being on the road, doing his spiel, but still keen on making a difference. He had just done a tour of Canadian cities, the Small Blue Dot project named after Carl Sagan's moving description of our planet seen from outer space as that tiny sign of life amid the darkness. They were asking the big cities to change their constitution to commit to preserving soil, air and water and many of them agreed. The constructive Suzuki is unbeatable.

His Foundation is also impressive. I saw it in Vancouver, filled with bright and beautiful people doing things to save the world (I say that in the nicest possible way). And for those who raise eyebrows at Suzuki's stellar income, may I confirm that his dollars do go to this worthy effort and its sheer size indicates that the Foundation must take a hefty set of dollars to run.

* * *

In the beginning, back in 1975, and for a couple of decades, we had old technology and plenty of reporters everywhere. The technology consisted of a Nagra reel-to-reel tape recorder weighing over 12 kilos, which I carried on my left shoulder, as result of which I list permanently to port and am getting treatment to have my posture corrected. The tapes weighed nearly as much, '5-inch' reels in boxes of which I carried a few and up to fifty when abroad. There was room in my case left for two underpants, a couple of shirts and a toothbrush. We took it for granted, but it was taxing. I remember an occasion when cab vouchers were no longer available in the ABC because money had run out yet again. I was due to interview ex-premier of New South Wales Neville Wran on his creation of the HECS tax for students so that they could pay back loans enabling them to attend university. I carried my 13 kilos-plus kit across Sydney, there and back — 7 kilometres. I was somewhat moist and shambolic on arriving at Wran's smart office. We knew each other well and he accepted the explanation for my disarray: 'F-----g ABC! F-----g federal government!'

Making all this lumpy technology work was also a performance. To send material from a distance you had to go to a public phone box, dismantle the speaking part on the handset, and connect wires from your Nagra. Then you played the recording in real time. Many an irate member of the public would knock on the window if we took too long. Freelance material would arrive on cassettes or those 7-inch boxes having been posted or sent in a bag from an ABC overseas office. Immediacy, as experienced today, was not possible, but urgent reports were done by phone direct. Now I will not consider phone-quality interviews as the sound is usually dire. Back then there was no choice.

As for those reporters, where are they now? If they were freelancers you had to commit to a regular commission, as no journalist can survive on $200 every few weeks. There was Bernard Mayes in San Francisco, a British one-time actor and someone keen on science. His Radio Shack tapes turned up for years, reliably and with delightful scripted reports. Then there was Doug Crawford in London, a cheeky chappie, but supremely professional. I have not heard from either of them for decades. There was Wendy Barnaby who started in Sweden when her husband Dr Frank Barnaby was appointed as director of the Stockholm Peace Research Institute. Wendy was Australian, born in Adelaide but with a *Downton Abbey* voice and intonation as is often displayed by Adelaideans, and I trained her by phone as she fiddled with a cassette recorder on her kitchen table. Wendy did many a superb feature for Religion on RN in later years, but has now retired from journalism.

It sounds almost quaint referring to these worthy hacks in faraway places sending in their parcelled items, but that was the nature of radio way back. Even TV depended on videos carried in planes and trains. Listeners seemed not to suffer from hearing about a 'breakthrough' one or two days late. We also got lots from our ABC offices: from Ellen Fanning when based in Washington DC, from Matt Peacock in New York (yes, we had an ABC office in NY for some reason) — and being 'Manager' for NY was virtually a diplomatic posting, with an apartment on the northern side of Central Park which had a view worth millions. Correspondents in Moscow (yes, John Lombard in a furry hat), in Papua New Guinea, in South East Asia — we were like a real grown-up radio station. Now — all gone. Our correspondents may have to do four or five feeds a day for our many networks and can barely manage that, let alone peel themselves away for the little *Science Show*. I hardly ever go near what remains of our overseas offices now — there's no point.

All this means that the nature of the program changes every year. Where once you heard journos, now you hear PhDs. Where there were ABC staff, you now hear writers (every writer wants to promote a book), friends of mine who work for the BBC or *New Scientist* or CBC and can spare time for a feature and anyone else who turns up and is not pitching for commercial products or a PR person flogging a client.

Meanwhile you have access to half a million new YouTube videos being uploaded every day. You have unlimited print items on the internet. You have podcasts of other broadcasters' programs and those of science journals (such as *Nature* and *Science*) as well as blogs from more science-associated bloggers (sometimes barmy) than you can ever meet in a lifetime, and all just seconds away on your devices.

Why on earth do you need a *Science Show* any more, or RN, or the ABC? Are we in the ash can of history, after these forty years?

12 The Future

Is there one?

Have you heard of MOOCs — Massive Open Online Courses? Many universities run them, from Harvard to the University of Queensland. Now. And they are free. You can see the world's greatest lecturers doing their party pieces on your screen in seconds. Millions of students are signed up for MOOCs. Many will drop out quickly, but untold thousands will stay the courses and get degrees.

Why would you need a *Science Show* when you can get a bonanza like that? Then there is TED, the talks on-line, vast choices on YouTube, podcasts from everywhere. Why would you need a little radio program, pray — or the ABC at all — when you have this cornucopia?

Are you a neophiliac, determined to cite new technology as the answer to everything, requiring universal change? If you analyse what most new technology in communication does with you, it is summed up by two elements: excess speed and fragmentation. Checking a fact (what is MOOC?) is done nicely on the internet, as is spelling a strange name or other factoid. In fact, the inaugural director of Balliol's internet institute, William Dutton, told me once that he realised a revolution was at hand when it took him more time to cross the room to look something up in a book than it did just to Google it at his desk. But an opus, a speech or a longer article needs attention *at leisure*. Skimming a thoughtful lecture is no use, unless you just want to name-drop, to say you've heard it. Ultimately, your choice is unchanged, despite the near miraculous nature of the new devices and the speed with which they fulfil your desires. But apart from easy access, you *still* need focus and reflection for a major article, a one-hour radio program or a talk by a major authority. You need undivided or *less divided* attention.

Fragmentation is something we all feel when confronting the infinity of choice and overwhelming distraction of what is e-available. It is why I refuse to own an iPhone or any other kind of mobile. I do not need any more messages. As it is, my workload has doubled because of the flood of communication and the incontinence thereof. I get into the office at 6 a.m. to cope with the deluge and work seven days a week to keep up (today is a Sunday!). The ABC is not interested in this. They say something heart-warming about a work–life balance but do nothing to make our desks more functional (we have no offices anymore) or staff assistance to give us greater ease. 'There is no money!' is the refrain we hear instead.

Facebook and Twitter (that annoying phrase now even beyond cliché) I also eschew, for the same reasons. If you can act on just twenty of the hundreds of email requests and demands that come every day you're doing well. Add the other lot (Tweets, etc.) and we're into nightmare territory. So, a *Science Show* or similar program aims to give you the *opposite* of fragmentation: a coherent guide to what's important plus a level of production that gives you a relaxed experience. It can help if you put aside the proper time for this longer program to go for a walk or run while listening, or clean out the shed or kitchen at the same time.

You find a similar story if you examine popular programs here and overseas, say *This American Life* (National Public Radio) or *Radiolab*, let alone the long series made by HBO, which have transformed TV viewing. One of the most popular podcasts on ABC Radio is Richard Fidler's *Conversations*, again, an hour long and with spectacular ratings. The same applies to LNL (*Late Night Live*) presented by Phillip Adams for 25 years and, again, with spectacular ratings. People can choose to make the time to listen — even young people, who have to train themselves *not* to be distracted after seven minutes. What really is the demand on their lives that is different from how ours was, apart from the twitch to nip from one device to the other to be told something utterly trivial about someone's shoes, lunch or bowel movement (though in my case, the latter would be important news)?

What I have just written applies to 2015 but it will equally apply in 2050 and beyond because we are still simple animals, human ones, with needs that rarely require lightning speed and unlimited choice. A person talking to another person, a story in a book, on screen or in your hand, a lesson to be learned for school, work or play — we are not yet robots with IT implants!

Yes, if I compare the convenience of the new technology with the vast clunky machines of thirty to forty years ago, life is transformed. An interview or feature recorded anywhere on Earth can be attached to an email and be with me in seconds. Even though I kid the ignorant that I never use computers or have never given up reel-to-reel recorders, I did pick them up as soon as they proved they were useful and efficient. We were among the very first to say 'yes' to podcasting. My interviews, maybe thirty or forty of them, can be on one small card inserted into a lightweight digital recorder. This weighs a tenth of a kilo, not 12 kilos. Such an improvement! We welcome what works. What we don't want is a replacement technology every few months.

But I still do my scripts on a typewriter. This is for two reasons: the first is that I am offended by the tonnage of new paper being used promiscuously wherever I look. ABC staff are basically environmental vandals like everyone else. Lights are left blazing over nights and weekends; computers remain on with fans blowing for days; and paper is used on one side with entire books being printed, then thrown away. My scripts are all on old press releases sent to others and found next to copiers or in bins (I never print them) or ABC memos, or even the blank side of magazine address sheets extracted from their plastic containers with my address on one side and nothing on the other. So my new paper bill since 1972 has been zero dollars.

The other reason I rely on the typewriter for a *Science Show* script, apart from the fact it never crashes and I have no interruption crises, is that I play the sound through the computer. My experience is that, if I use the computer for both, I get the confusing effect similar to driving and using a phone at the same time. I often hear something clunky in a colleague's on-air script and can usually tell that they didn't notice anything amiss because both sound and words were jostling on the same system. Besides, typing on paper seems to encourage brevity. My scripts are famously succinct.

What about the dearth of reporters? Well, we take one step nearer the source material. Instead of a middle person telling you about a brilliant PhD scheme, we have the PhD in person doing this. Which brings me to the first example of the revolution underway, as it's been for three or four years.

Communication of science by university people in Australia has improved by a factor of four. It is now uniformly wonderful. Whereas about ten years ago a lecturer or a PhD would stand mumbling in front of an incomprehensible PowerPoint, now they speak *to you* — instead of to the unappreciative screen with its dot points — and sound both informed and enthusiastic. This revolution was brought about by an initiative from the University of Queensland: the Three-Minute Thesis Competition. Niall Byrne's Fresh Science scheme did something similar. He had speakers light a sparkler and describe their research in the minutes before their fingers got burned. Nowadays, they time three minutes with a stopwatch instead (Health and Safety strikes again). And, over the years, the young PhD people have learned from each other and previous combatants. Gone is the sing-song style of the few paragraphs learned by heart and recited; gone is the tendency to regurgitate facts and numbers at length; gone is the set of graphs: only one picture is allowed. Instead, the youngsters stand there and *talk* to you, with directness and spirit.

This has two rewards for them. The first is that putting their work into plain language makes them understand their own science better. The second is that they become equipped with articulacy essential for a life in research, or even one in business. Being able to speak in a compelling way is vitally important in most careers.

I have been a judge many times for this competition, at University of New South Wales, Australian National University, University of Queensland, University of Western Sydney, University of Sydney, and at Famelab, an international competition fostered by the British Council and the Cheltenham Science Festival, attracting hundreds of entrants. There too the standard is superb.

A real test of this standard was the Top 5 Under 40 competition (five scientists under forty years of age), sponsored at the ABC by UNSW, to celebrate forty years of *The Science Show*, in which hundreds of young scientists were invited to audition as speakers with the winning five joining us at the ABC (as did the interns before in the McGauran scheme — we have had a number of phases of encouraging internship), to learn about media culture and make programs for us. As a judge of the twelve finalists, I was staggered by the high level of achievement. Then I was told that the last fifty to one hundred were almost as good and that it was heartbreaking saying no to so many. Watching them talk about their research and ambitions for an ABC stint made me think, 'I could not have competed with these people: I'd have been back out on the street after the first cull.'

It was an incredible experience. What had happened to all the muttering misfits we had a generation ago? If this is the new standard, we could run twice as many science programs: the talent out there is an inspiration. And all this from practice and encouragement, in just a handful of years.

This is one reason I so regularly feature the work of PhD students on *The Science Show*. Almost any example of their research could be applied to innovation and therefore wealth creation. They are the shining promise for tomorrow. They must not experience the rejection and insult so many of our hardworking Australians do: 'Sorry, no money; no vacancies; no chance.' Then, after fifteen years of intense study, a PhD graduate has to drive a cab or try something overseas — or just give up. I found my degree to be the hardest I'd ever worked in my life — and I did not even do a PhD.

So let me not hear, ever again, some out-of-date bystander remark that scientists are inarticulate, boring speakers, and bad talent. Not anymore in Oz! They have been transformed on campus by their own efforts, and by regular appearances on our programs — and not just Science Unit shows or *Catalyst*. The number of ABC (and SBS) programs giving a place for science material is surprising and large — *AM, PM, Mornings* (Linda Mottram and Jon Faine), Tony Delroy on *Nightlife* (ABC Local) — every rural and regional ABC Radio show I've ever heard has a go, as do *Bush Telegraph* (RIP!), *ABC Rural, Saturday Extra, Future Tense, Drive*. The list goes on (with ABC TV too — *7.30, Four Corners*).

And so to the second revolution: Citizen Science.

If speaking well has become the rule in science in the past few years, so has Citizen Science in its modern manifestation through the new technology. This participation from members of the public has had, in bird observation and botany, a very long history, going back to the nineteenth century. Gentlemen and ladies at that time went wandering along hedgerows and copses spotting unusual species and listening for the first cuckoo. It was a respectable occupation for those with spare time and was the basis for precious discoveries. Mary Anning, who lived in the west of England, helped invent modern palaeontology by strolling along cliffs and seashores looking for fossils.

In the twenty-first century the field has become much more demotic and vastly populated. One of the best-known modern manifestations is Galaxy Zoo, run by Chris Lintott and friends (he took over from Patrick Moore presenting *The Sky At Night* on BBC TV). It now has well over a million citizen supporters.

It began when a researcher looking at assorted galaxies realised that there were so many to classify, in trying to get more clues about the growth of the universe, that it would probably take 370 years for him to finish his task. So he asked for help online. All you had to do was sort galaxies into two or three classes and report on numbers found. The website melted; so many people, from age six to over ninety, wanted to help. And they did so. They also got feedback. This is vitally important because if you treat people like slaves of no account they just go away. Another bonus was that the amateurs had time to look for other things as well as simply sorting, and they took delight in informing the astronomers what they had found. They, in turn, were enraptured. More discoveries!

Soon there were citizen scientists covering innumerable unexpected fields, from cancer to fish migration. At the American Association for the Advancement of Science meeting in San Jose, California in 2015, I blundered into their national conference with delegates from 700 different fields. I talked to one of the organisers, Jennifer Shirk, of the Citizen Science Association in the USA, about this incredible proliferation.

> We are growing out of a large tradition of natural history research and amateur involvement in everything from ornithology to butterfly-watching, and now are extending into areas such as cancer research and astronomy as well also very local-scale water-quality monitoring projects.
>
> People are most familiar with what's around them in their own areas and are also very astute at doing things

like analysing digital images, much better than computer algorithms can do.

We have some folks here who are leading in water-quality monitoring and analysis, for example, Kris Stepenuck, who is at the University of Wisconsin. She does work to support communities in analysing and understanding the health of their local ecosystems, specifically water related, and she supports that work around the US. Others would include people who are doing marine research. There are folks here from the National Institutes of Health, again looking at cancer, looking at the microbiome, the microorganisms that live on us and around us.

There's a program out of North Carolina State and the Natural History Museum in North Carolina [that is] inviting people to send in swabs of their homes, and in some cases their bodies, [in] a program called Bellybutton Biodiversity.

RW: Bellybutton Biodiversity! And you've got citizen science people taking those kinds of samples?

Jennifer Shirk: Not just taking those samples but helping scientists to understand what those samples might mean, so making hypotheses about why there is such a difference in the microbiome from home to home and from individual to individual. People are aware of their own surroundings and they can help scientists better understand how to make sense of what these data are showing.

RW: Now, what do you have to do, and I ask you as an organiser of citizen science, to give feedback to people who otherwise might just be used as a kind of slave labour, and never hear back from scientists? How do you keep their interests and their evolution as amateur scientists?

JS: That feedback loop is really critical, and many organisations work to not just give feedback but to provide ways for the public to be more closely involved in the research process. Again, like Bellybutton Biodiversity and inviting people in to understanding what the data mean. It's a hard thing for scientists to do, and that's one of the things that citizen science can really provide, an opportunity to make those closer connections with the public, to not just share but to listen and to learn.

Australia may soon have a movement on the same scale. I have already mentioned, in a previous chapter, the cray-fishers from the Abrolhos Islands in Western Australia. We have many Australians linked to the international sites too, including Galaxy Zoo.

Here are some of the comments from those in USA, UK and elsewhere on the force and promise of this movement.

Chris Gillies, Earthwatch

Citizen science is not new at all. It was the way of science right up until about the 1900s when science became a formal profession. Before that, citizen science was really in the realm of amateur scientists, but citizen science has not been new to Earthwatch. We have certainly been doing it since 1971, and nobody else was really doing this week-long intensive engagement where you can get out there, you can jump the other side of the fence, if you like, you're out of the tour bus and you are working alongside scientists in the field. You're actually touching, feeling and engaging with the wildlife and with nature, and you are also assisting scientists while doing that. So you really get that inspiring opportunity to be out in the field and to work with scientists, but at the same time it's lots of fun and you get to explore new places.

Earthwatch supports more than ninety research expeditions globally, and 60 per cent of those are open to the public. Our traditional research expeditions that anybody can join range from working on the Arctic edge investigating the impact of climate change on permafrost, you could be tracking chimps through the trees of Uganda, or doing some shark conservation in Belize. Closer to home, here in

Australia, you could be looking at manta rays on the Great Barrier Reef or indeed investigating the impacts of climate change on Australia's wet tropics up in the rainforests of northern Queensland. You are in small teams of eight to twelve individuals and a number of different researchers, and you're living and breathing and acting just like the scientists that you are working alongside.

Scientists can't be everywhere. We can't be sending out thousands and thousands of scientists across Australia, across the globe. We really rely on the community; we rely on citizen scientists to join scientists in collecting some of those observations. Volunteers are really a critical component now of conducting science, and they add a lot of capacity to the way that science is done today.

David Karoly, Professor of Meteorology, University of Melbourne

What we're asking people to do is to make available their home computer to run a simulation of the global weather and climate, and the Australian and New Zealand weather and climate. We are really asking them to allow us to run an application, Weather@home ANZ, on their home computer, and it sends back their output. The daily weather variations for Australia and New Zealand are stored at a data server actually housed at the University of Tasmania. There are somewhere between 20,000 and 30,000 people who have already signed up to this project. We already have 18,000 repeated simulations of 2013.

Chris Lintott, Citizen Science Project Lead, Adler Planetarium, Chicago, USA

In the last five to ten years we've seen a shift to projects where the professionals provide the data, thanks to large surveys, and we ask for volunteer help to sort through that data and turn it into knowledge. And we are celebrating today a milestone, which we reached just within the last twenty-four hours, which is the (citizen science web portal) Zooniverse's one millionth registered volunteer.

Those volunteers have discovered planets around other stars, including the first planet in a four-star system, and they've helped cancer scientists in the UK deal with the problem of pathology, of having to look through millions of cell samples to detect reactions to drugs. And they've expanded themselves into problems of climate change, looking at historical data, rescuing data from ships' logs, and also to the humanities where we have projects looking at the history of World War I, for example.

We have a huge breadth of effort, and what we are interested in is how to grow that into a very sustainable citizen science community that can be of use to thousands rather than tens of scientists. [We want to] examine the different ways in which we can align the motivations of our volunteers, who typically want to help with research, with those scientists who want to get their work done while maintaining a sense of real participation.

Caren Cooper, Citizen Science Programs, Cornell University, Ithaca, NY, USA

I'm at the Cornell Lab of Ornithology, which is a hub of a lot of bird-related citizen science. Citizen science happens online, it happens on the ground, it really is a term used to describe a whole array of genuine research activities that the public can take part in. You could look at almost any scientific discipline, and if you look deep enough and carefully enough you're going to see some aspects of citizen science happening. You can see it with the people here today — I mean, an astronomer, a neuroscientist and an ornithologist walk into a session together ... It sounds like a good joke.

Here we have *millions* of people around the world in regular contact with top working scientists and getting encouragement and feedback from them. Can it counter the drop away from science in some parts of formal education in Australia and the paucity of jobs? Add this movement of amateurs to the impressive articulacy now abundant on Australian science campuses and you have a combination that could take off and really change the culture. No more apathy; no proud ignorance, but a scientifically minded nation with plenty of essential investigation to get on with. Am I kidding myself? It could, as the adolescents keep saying, be 'mega'.

What's the alternative? Australia keeps trying to fulfil Lee Kuan Yew's prognostication, either because it can't help it or is being undermined by the kings of 'the Lucky Country'. But we really do risk becoming the 'white trash of Asia'. This year and next year are pivotal. All of which leads us, briefly, to science policy.

* * *

The science minister who succeeded Barry Jones under Prime Minister Hawke was Ross Free, a former science teacher. He did what every new minister likes to do: commission a survey or an inquiry. It arrived bound in shiny plastic covers and I dubbed it straightaway the 'Nerds and Losers' survey.

The investigators had been to schools to ask various ages of pupil what they thought of those good at science. The youngsters quickly thought of the myopic misfits in the classroom with their fiddly fingers and querulous voices, as well as the infuriating habit they had of answering questions whose incomprehensibility was matched by the arcane answers. They described these poor possums as 'nerds' and 'losers'.

The double bind: those keen on science were seen as both twerps *and* bound to fail in their quest for glory. I was supremely irritated by the report. As usual it told us nothing we didn't know already and made anyone inclined to take some sort of action become catatonic or want to spend the next week in the pub, or both.

How many times have we heard that Australia is short of engineers, mathematicians, and teachers, people who can walk and chew gum leaves at the same time? How often have the solutions been offered? Make science teachers' salaries objects of envy, give the profession the status it had when I arrived in Australia (I was turned down as not properly qualified — quite right!); make science jobs secure instead of on/off like a prostate sufferer's dangly bits; tell the scientists once more, as Barry Jones did, not to be wimps; and tell the politicians not to stand at posh dinners once a year and tell scientists they are God's gift, only to ignore them for the duration.

(At that Ross Free meeting, there was a call to write a soap opera for TV with a scientific theme. I told them, helpfully, that drama on telly was *the* most expensive form there is, but that I'd write something for *one* person on radio instead. It would be a take-off of Adrian Mole. I immediately got to work and wrote some scripts, which were delivered to Glebe where Andrew Denton read them dutifully into a mike. The series ran on *The Science Show* for a couple of months.)

Chris Evans, as science minister, briefly, under Labor, did not praise the boffins at the posh dinner in The House in Canberra on the day that Science Meets Parliament (an annual event). He wasn't going to echo the standard line that they are all treasures and walk on water, then leave the room and do nothing. He told them instead to stop thinking about themselves as special and yet singled out for neglect. He said, in fact, that they *are* considered when appropriate, like everyone else is; that there are many elements to policies, of which theirs is only a small part; and they should back off and get their acts together.

It was deeply embarrassing. The hall, once more, knew all that. They wanted to hear how best to cut through, not just for self-preservation but for the needs of the nation and recognition of first-class work. Instead, everyone was left with a sour taste and Chris Evans promptly retired back to the West. *Plus ça change.*

Are there examples in other nations of how it can be done effectively? Well, plenty. Israel invests 4.3 per cent of GDP in innovation and has the most start-up companies per head of population in the world. That is twice the investment Australia can manage. Germany, Sweden, Norway and even the USA do much better, by making resources secure, forgiving failure and having a farsighted view of the nation's needs. Most of the time our governments can't organise a shortish trip to the lavatory. Perhaps 2016 will see a better approach.

I know it is hard to effect change in office, as Barack Obama's sad regime has proved until very recently. But in Australia we have those famous envelopes stuffed with cash (causing *seven* MPs in NSW to have to resign), deals done with dubious mates, decision makers who can't decide, rhetoric and 'zingers' instead of confidently mounted ideas, and policy conjured from bewildered focus groups instead of thoughtful expert analysts who've been to where people work and live and listened to them properly. Peter McGauran, twice science minister under John Howard, was a success because he did the leg work and saw what the nation needed.

My friend at the Wellcome Trust in the UK, Mark Henderson, used to write for *The Times* of London as a science reporter, despite his reading history at Balliol. A couple of years ago he brought out a book called (unfetchingly), *The Geeks' Manifesto*. All the policy you need is there: how to assess your science's merit; the perils of unpredictable funds; the way to obtain science teachers quickly (hire retirees); how much good research will save the nation (untold billions); and how wealth can be created to make a difference here and abroad. The book will take a day and a half to read — and it will be worth every minute.

Mark Henderson: Without [the] experience of science, it means that when politicians or ministers come to manage science, they often do it badly. You'll have regulations that are brought in with very well-meaning purposes, to protect patients undergoing clinical trials or whatever, that end up creating a bureaucracy that interferes with the ability to do those trials in the first place, simply because they're not framed with the necessary experience.

And then the other factor is that without this understanding of something that the great astronomer and broadcaster Carl Sagan put brilliantly — that science is more than a body of knowledge, it's also a way of thinking — without that understanding, which I do think too few politicians have, most politicians (like most laypeople in general) tend to think of science as a bunch of facts rather than as an approach to acquiring and improving knowledge. But without that understanding there's too little thinking about how that scientific approach could contribute to creating better policy, producing evidence through some of the tools of science. Randomised control trials, for example, that we use in medicine all the time, have huge potential to be used in other realms of public policy, education or criminal justice, to really show whether different interventions work or not, and we don't do that enough.

RW: The ministers of course want an answer by tomorrow or next week, whereas science takes time.

MH: That's true, and there are of course times when it's essential for politicians to act without much evidence. You can think of all sorts of examples where something has to be done straightaway. A good example would have been when the [eruption at the] Icelandic volcano Eyjafjallajökull took place, ministers had to take decisions about whether to close air space, with relatively little evidence about the effect of volcanic ash on jet engines. But, first of all, there was a reason there was so little evidence. Actually the British Geological

Survey had gone to ministers a few years beforehand and said, 'Do you know what — Iceland is, kind of, quite an active volcanic area, and it also stands slap in the middle of the major transatlantic air route. Shouldn't we do a bit of risk assessment?' And they were simply ignored. So there was a reason there was no evidence there.

But bearing in mind that of course ministers have to take decisions like that all the time and can't always wait for evidence to be gathered, we don't do nearly enough to then evaluate those policy decisions as if they were experiments because, of course, they are experiments. Every policy that's introduced is an experiment, but the problem is we very rarely bother to collect any data from those experiments so we can't learn from them. There is a disincentive even to doing that because if you do capture the data properly and do some rigorous evaluation, it may well show that this policy, which a minister put great political capital by, didn't work. And so it's almost easier to commission some kind of half-baked analysis that won't prove you right but it won't prove you wrong either.

* * *

Back to the science itself and what we may expect to see in the next few years. Can Australia play an important role in what's to come?

1. Down come the boundaries Nature is not divided, so why should its study be? In the past week I have interviewed a mathematician who studies genomes, chemists who do marine research, astrophysicists who investigate cancer and epidemiologists who look down on Earth from satellites. How do you study human diseases by looking down to Earth? Trust me — it happens at the ANU. Combining disciplines can reveal so much more. This revolution is well under way and Australia is already involved. May we stay that way.

2. Materials Graphene is a new wonder material that won two guys in Manchester the Nobel Prize for Physics. They obtained it by peeling off Sellotape that had been placed on graphite. The one-layered atomic sheet has stupendous strength and can be used for everything from ultrasensitive condoms to powerful conductors and vast sails for long-distance space vehicles. Monash is one graphene HQ and there are others in Oz such as at the ANU. But the material's story could be transformative. In Geelong, as the car industry crumbles, a new field for building vehicles (electric driverless cars?) and much more, could be the basis of new industries, as Jane den Hollander, Deakin University's vice chancellor insists.

3. Brains We do superb work on nerves here. The teams in Western Australia, led formerly by Dr Lyn Beazley, showed how you could reconnect severed optic nerves in amphibians by applying growth factors. Making the blind see again! For lizards, who have nervous systems more like ours, it was necessary to also train them after the repair was made. (It was fascinating that 'rewiring' was not enough and practice as, say, catching prey was needed. This confirms experience with human experiments involving technology such as the bionic ear: you need to train to get results, after the reconnection.) And, in Brisbane, the University of Queensland has superb imaging techniques enabling you to see which parts of the brain are active as you do things *and* they are excellent at using genomes to tell what may be going wrong and when. And then there's the Florey Institute in Melbourne, a famous brain study centre.

4. Immunology Why is this field so superb in Australia? One wag said it's because you need fewer funds — cheaper apparatus — but there's more to it than that. It's leadership. Mac Burnet was a genius and when you have one it's likely followers will flock. And they did, not least at the Walter and Eliza Hall Institute.

5. Genetics I know she lives in San Francisco, but Elizabeth Blackburn did come from Tasmania and Melbourne University, *and* she won a Nobel Prize. Her treasures are the caps at the end of chromosomes, called telomeres. They control your wellbeing, if they remain long; if shortened, your resistance to disease will fall, usually because of stress. Relax, rest, and they grow again. This field is about to explode, in Australia as well.

6. Geology It was shocking about twenty years ago when our universities closed half the geology departments in Australia. What about mining and related industries, I wondered, don't they count? The ones that remain, in Western Australia, Melbourne and Newcastle, have strong related fields, including engineering.

7. Chemistry Two Melbourne chemists working on polymers are often mentioned as potential Nobel Prize winners: Ezio Rizzardo (CSIRO) and David Solomon (University of Melbourne). Then there is the team from the University of Sydney, led by Tom Maschmeyer, doing so much you get giddy (batteries, fuels). And then there's the Green Chemistry clan at Monash: over a hundred scientists and engineers with exciting ambition such as using coal as a source for a chemical industry instead of for burning.

8. Cities The team led by Professor Peter Newman at Curtin University in Perth combine transport, energy, housing and waste — all applied to cities of the twenty-first century and beyond, making life more pleasant, relaxed and clean.

9. Quantum computing I have already mentioned Professor Michelle Simmons's staggering achievements in this field. She is manipulating single atoms as if they are cherries *and* having a *Nature* journal at her command to give the field profile. And others elsewhere are also doing well.

10. Maths and statistics Both fields are also excellent, but you have to accept those from elsewhere to give the accolade. Terry Speed from Melbourne won the Prime Minister's Science Prize for using statistics to measure the occurrence of certain ailments so that epidemiology and resources could be marshalled.

11. 3D printing at the University of Wollongong You've heard of 3D printing of the 'conventional' sort, such as tools or engine components, but Wollongong's Innovations Institute is printing muscle and nerves. No, I couldn't take it in either! Now they are printing minute scaffolding that can carry the stem cells they seed to bridge the gap in the tissues being repaired. The Institute's reputation, built on intelligent clothes such as an athlete's bra that expands and tightens with the athlete's movements, was set years ago. Now they have more than a hundred researchers working on the seemingly impossible.

12. Big data Yes, the SKA had to turn up somewhere on this list. The Square Kilometre Array will be going up from 2016 but the computer systems are already growing in Perth. To process the amount of data pouring in they will need to be equivalent to all the computer power in the world today *combined* (yes, this is worth repeating). Will the internet cope? *New Scientist* has doubts. New internet carriers, some made of light, will have to help with the expansion. And handling Big Data is not a doddle, but get it right and you can apply it to mining operations (important in Western Australia) and much else.

I'll stop at twelve. It could be fifty, but you get the picture. Some areas are neglected and we'll come to those. But one question: is Australia simply too big for some fields so that researchers will inevitably become stretched too far?

13 Is Australia Too Big?
Or, why do we sometimes think small?

Look at the map: Australia is vast, but it often looks at itself as a junior league player — unlike New Zealand always playing David against a range of Goliaths. Thirty-eight per cent of the continent is tropical, yet most Australians think the Top End consists of a couple of towns (Darwin and Cairns) and maybe a bit of scrub and wetlands all around. Cut out the crocs and the cyclones and what is there?

More than a third of Australia counts enormously. All those minerals, fossils, indigenous people living what kinds of lives? Could it be that too many of us live in a handful of capital cities on the coast and we rarely see the extent of this huge nation? Tasmania is larger than Austria or the Czech Republic and the Kimberley is three times the size of England and twice that of Victoria. And the Top End is like a different nation. The weather up there is so powerful that Cyclone Tracy killed a city. There are creatures apart from crocs. Just think something that used to happen in the nineteenth century: expeditions set out from Europe, led by gentlemen to explore regions unknown to research, apart from being an undifferentiated blob on a big map. Well, such an expedition was led by geographer Professor Andrew Goudie from Oxford twenty years ago to the Kimberley in Western Australia. Yes, to modern Australia. And they found scores of new species — not surprising when you think of its scale. And now Vickie Laurie has just published a superb book, *The Southwest*, on another 'discovery' — of a 'Hot Spot' further south where biodiversity flourishes at an unimagined level. But back to the north...

Take the next step and link our tropics to the rest and you suddenly realise how gigantic the prospects will be. Professor Sandra Harding has done this, in 2014, launching an international investigation of what Aristotle once dismissed as a part of the world in which no proper person should live (he favoured temperate zones). But the figures, now in, are staggering. Soon 67 per cent of the world's children will be inhabitants of the tropics. This is how Professor Harding, vice chancellor of James Cook University in Northern Queensland was able to enlist Aung San Suu Kyi to launch the report, *The State of the Tropics*. Think beyond boundaries: the future will be utterly different.

Western Australia is five times the size of Texas. Now add the south — yes, 'down' to Tasmania. Do you realise that the latitude of Hobart actually matches that of Rome in the north? But then you have the Southern Ocean and our part of Antarctica — huge distances for a population of 23 million to cope with, surely. But Norway and Sweden manage, and Canada sometimes leaves us looking silly, as we'll see.

So what disciplines need to be managed on a continental scale? Well, there are Aboriginal studies, space industries, marine science, astronomy, and engineering of large structures, such as railway lines, that meet and match. Then there is ecology, though you can and must often concentrate on a small patch, and do 'blitzes' when twenty to forty mixed specialists descend on a hotspot or hillside and measure *everything*. This happens now and one of the hotspots examined was in the south-western part of Western Australia and their reward was innumerable new species and special sightings (as we broadcast in *The Science Show*) — so you don't have to spread yourself thin to get significant results.

Yes, policy recommends that you concentrate on what you are good at, admit you can't do everything, keep some areas ticking over in a healthy state, and then come back to them when you can do a proper job. Don't just abandon whole slabs of activity indefinitely.

But we have, in the past few years, had real embarrassments exposing our paucity of commitment. Take fossils with backbones. In the 1980s when I was president of the Australian Museum, one day when it was very hot (45 degrees Celsius?) I chaired a board meeting (The Trust) in shorts and sandals. I looked around the table and saw my fellow trustees, mostly captains of industry, all wearing dark suits and ties. 'Why?' I asked. 'You live in a hot country; why dress as if you are in Edinburgh in mid-winter?' My deputy, head of an oil company as I recall, explained sweetly that the Governor of New South Wales, a retired admiral, was coming to see us at 4 p.m. and a little decorum was thought necessary.

I gulped, and then had an inspiration. I leapt next door and asked the secretary to call Dr Alex Ritchie, our dedicated vertebrate fossil man. I'd heard him on the news at midday announcing a stupendous find of rare Devonian fish near Canowindra in New South Wales — tonnes of the creatures. 'Bring a trolley of them to the boardroom at precisely 4 p.m.', I instructed Alex, and he did so.

At 4 p.m. the Governor appeared, gave my shorts a mere glance as, just then, the fossils appeared — and he'd heard about them via the ABC. He was hugely excited. All thought of my wanton informality vanished. We had won a supporter.

The point of this story is that the Museum's exhibition of Chinese dinosaurs a couple of years ago had no in-house palaeontologist to describe the treasures. The letting-go of staff had reached a point where our oldest and noblest museum did not even have a person aboard who could spruik *T. Rex* and its relations. They had to enlist an American from Scotland on the *phone* to do the honours. Much the same is the case in

Queensland. As Scott Hocknull and his volunteers unearth some of the biggest dinos ever found up in the north, there are hardly any professional fossil experts to work on the specimens. Meanwhile Clive Palmer spent squillions on his mechanical monsters in his theme park — and, yes, I did ask him to make a contribution to the Queensland Museum in this field — and failed.

I became worried about Australia's tendency to think small when I arrived here in 1964. I knew only one person in the southern hemisphere when I arrived: Jane Graves, with whom I went to parties in London in the late fifties and early sixties. (She is now my sort of mother-in-law, having before been married to the former chairman of Macquarie Bank, David Clarke). Then there were the Holmeses. They were part of the Goldacre family (Michael is my oldest friend, the professor of public health at Oxford and father of the famous physician and science writer Ben). They made me promise I would go to Young, New South Wales, for Christmas and join the Reverend Holmes, who was the rector at what I presumed was the Anglican Church — I am vague about such matters — but he was the grandson of Sir Henry Parkes, Father of Federation and once premier of New South Wales.

So I got in touch, scored an invitation and, on the Christmas Eve of 1964 set out to Central Station having already bought a ticket. I was disconcerted by the crowds but assumed all would be serene when the train set off. I found my carriage. It was as crowded as the Indian trains I would ride on two years later. I was shocked, having been used to European trains that were smooth and comfortable and which travelled at high speeds, near 200 kilometres per hour. I was stuck in the corridor. You couldn't even get near a seat.

The train, late, shuffled off. It was barely above walking speed. This lasted twenty minutes. I remarked to the young fellow standing next to me, 'At this rate we won't be in Young before midnight!' 'Oh, no. No. No,' he replied, amused. 'We get to Young at 4 a.m.' I was totally floored. This was not even nineteenth century speed. But I stood the whole way and, despite being only twenty years of age, was pie-eyed when I got to Young station to be greeted by the priest wearing a straw cowboy hat above his dog collar. He was unsurprised by my railway horror story, taking it as given. Whenever I asked locals why their trains were not even up to hopeless Third-World standards they simply shrugged and implied that Australia didn't do flash speedy stuff but rubbed along in that friendly drongo way as if ambition were an embarrassment.

Trains now, which I catch to the South Coast of New South Wales, are nearly as bad. They needn't be. Go west if you want an example of what could be done, led by Professor Peter Newman of Curtin University who had an office next to West Australian premier Geoff Gallop when they were designing railway lines to serve Perth. If you have a train that can go faster than commuter's cars, said Professor Newman, people will opt instead for the convenience of modern public transport. And they have. If you want an international example of how to get trains linked with other efficient systems of transport, go to Zurich or Vienna.

Here's another example of modern, flexible thinking, this time from Lapland. In the 1980s Barry Jones introduced me to Kerstin Niblaeus, a young woman with a PhD in quantum chemistry who was a science adviser to the PM of Sweden, Ingvar Carlsson. I was on a trip to Sweden and looked her up. She made me an appointment to see the PM to talk about some of their remote mining communities and how they could move to completely different industries as the iron ore ran out or the drilling became too expensive.

Off I went to see Mr Carlsson. My son, then fourteen, embarrassed me by refusing to come and set off for Stockholm's Hard Rock Cafe instead. The PM told me about Kiruna, inside the Arctic Circle, the northernmost town in Sweden, where miners had been working for years in harsh conditions you could barely imagine. Niblaeus put her fine mind to the task, consulted, and then announced a solution — which I actually saw on a later visit up there in 1991, and gobsmacking it was too.

What do you do at the top of the Earth with retrained engineers, miners and other people from the threatened workforce? Why, build a space station, of course. And there they were launching rockets and balloons for the Japanese and other nations. The community was thriving. Compare that, I thought, to the FIFO (fly in, fly out), closures and other disasters we have in similar areas around Australia. What could the application of a little imagination, Swedish style, do for us instead?

My own reporting of related topics requiring travel within Australia has been severely limited for decades. I must rely on other organisations' airfares to get me around (in a kind of haphazard fashion) to give speeches. Nobody in Broome or Broken Hill has ever been in touch, so a planned, extensive set of reports for *The Science Show* on cave paintings in the Kimberley (where I've never been) or engineering in the Silver City, or anything from Antarctica other than third-hand, or on the biology of the Red Centre, I must leave to others.

The Swedes, by the way, paid for me to visit Scandinavia.

There are clever answers to the problem of size and we are already demonstrating their potential. Precincts. If you have a centre doing well and expanding, build on it. This happened at Macquarie University as the audio revolution took off with the bionic ear (Graham Clarke, by the way, struggled to get funds in the beginning). Now there are many lines of R&D in the field of audiology, and the Macquarie precinct expands — and one day may even claim in the bionic eye as a success.

Wollongong has 3D printing, intelligent clothes and other innovations; Geelong, through Deakin University, has the carbon material and wiring precinct; and Perth is creating its potential monster through big data. If you have effective precincts — like in Dunedin, New Zealand, the southernmost university in the world — it's as if you are in an environment as dense as MIT's or Oxford's. And the world will seek you out.

Precincts, blitzes, small-scale successes like the trains in Perth, can override the tyranny of distance. We are beginning to realise this, but not enough. The overwhelming mission statement from governments here has been 'save money'. Even in defence science, I've heard this pathetic refrain: cut the labs, we can buy the knowledge from overseas. What are our leaders doing in the Big Country? Thinking small. It doesn't work.

* * *

So where is our space industry?

They can build such an industry in Lapland but Australia seems to have given up, as Professor Mark Dodgson remarked in Chapter Five. Too big a country? Well, the Canadians do well with applications such as their famous mobile space arm (Canadarm) gracing many a NASA station or vehicle. And Chris Hadfield, the astronaut superstar, actually led a team in the International Space Station.

Les Field, who represents the Australian Academy of Science on policy, says there are three clear reasons why a space industry is vital for this country. The first is military: think terrorists and early-warning systems. This is obvious. The second is remote sensing for fire, agriculture, forests and even health. There is a project at the Australian National University that takes images from satellites and can tell from them which major diseases may be spreading here. Waiting for the French satellite to come around in time and being allowed to use its services is, frankly, not good enough. We need our own. And then there is communications.

We were, I'm told, in our industrial heyday, the fourth in the world to launch a satellite. Aspects of Google Earth were developed in Sydney by an American living here, namely Lars Rasmussen. Such is possible. In May 2015, another embarrassing venture exposed the limits of our enterprise. Harrison Steel, at the University of Sydney, where he's an undergraduate (!), is designing both a rocket and a satellite with his fellow students. Where will this enterprise find an outlet with no Australian space effort to snap them up? When I asked how they can do so spending hundreds instead of millions of dollars he smiled saying that they look for less costly approaches such as 3D printing their components! Professor Field tells me there are three public servants somewhere in the federal Department of Industry who have a vestigial responsibility for the space industry but anybody proudly bearing that name is long gone. So where will young Mr (Dr?) Steel go when he graduates and needs a profession? Overseas?

The Defence Science and Technology Organisation in Canberra and South Australia do have plans for several miniature satellite components to be launched soon. They insist that small *is* now beautiful. But this is vestigial work compared to our space industry ambitions in the 1980s.

It is yet again a problem of investment. As I write, the figures for investment in enterprises in this country are said to be dire and promise to get much worse. What could we do to emulate the Scandinavians? They too have had mines and know about doing what you've always done, then doing it again — dig-dig, carry-carry, sell-sell OS — then repeat a million times. Is it easier than *thinking* of a space-age industry? (Look at Mark Dodgson's comments on innovation. It is all there.)

14 Terra Nullius
How people first came to Oz ... and NZ

When I arrived in Australia in 1964 it was accepted that the indigenous people had been here for barely 4000 years. It was instructive, that as soon as I set foot on the soil of Sydney and was given an address, I had full right to reside, vote, pay taxes and get all the benefits of the Australian citizen. At that time, Aboriginal people were not even on the census lists, let alone allowed to participate in democracy. I have not changed my status since then and still have a European passport and do not hold Australian citizenship. What's the point? My family is Welsh and we're still trying to get rid of the Anglo–Saxons in the UK, not to mention the German–Greek-born royalty. And here I have the full privileges of the citizen — because in 1964, like you, I was a British subject. Unless you're Indigenous, of course.

When I started my broadcasting career, Aborigines were suddenly discovered to have been here tens of thousands of years. Now, it's 50,000 years, possibly 60,000! This is an almost incredible turnaround. From (relatively) new arrivals, the First Australians were being recognised as having the longest continuous culture of any people on Earth.

It was, for several years, one of *the* science stories. Even in the last stages of the Fraser government it seemed that every week in Tasmania Dr Rhys Jones and his archaeological mates were discovering a new cave with human remains or debris metres deep. Some assumed it was a put-up job and the sites had been found before yet kept secret, only now being released to embarrass the conservative governments who were keen to flood the Franklin River and exploit surrounding terrain.

But, no, it was all legit. The reason the fossil hunters were finding anthropological gold was that they were looking for it. Up until then not much had happened. Now they were finding evidence of the most southerly passage of humans in early history and signs of a declining technology. Was it true that your culture becomes less sophisticated as your numbers decline, as seemed to have happened as Tasmania lost its connection to the mainland? And that then the cycle reinforces itself because people have fewer skills and tools to make life easier?

Rhys Jones and his mate Jim Allen even suggested that Tasmania had been cut off completely from the mainland for 12,000 years so the southern people languished with no input from elsewhere. They were left with the smallest array of tools and gradually declined. The first they saw of other humanity after all this time was of the Europeans, the most developed and rapacious crowd on Earth, who quickly crushed them. All this was featured by Tom Haydon, a member of our ABC Science Unit, in his famous and influential film *The Last Tasmanian*.

> **Rhys Jones:** Twelve thousand years ago, when the lonely Tasmanian destiny began, all the world were hunters with many people's ancestors hunting reindeer and painting caves in the south-west of France. Looking at it from this perspective, we seem to have two histories: one the closed world (of Tasmania) and the other the open one with the same time scale, but within that time scale the development of agriculture, the development of industry and eventually expansion into all corners of Earth. The two worlds again met with devastating consequences for the Tasmanian society, which in a terrifying two or three generations was

utterly extirpated with almost all memory of it gone. The land was occupied, the culture destroyed, Truganini became a trophy; even genetically, the people were almost lost. Yet a small number of Tasmanian women became part founders of a population, [and] some of the descendants now again identify as Aborigines — [but] language, religion, knowledge about the land and the memory of the old things have gone.

I was like the historian archaeologist wandering through the landscape. I think the title *The Last Tasmanian* — which was meant to be an extreme left-wing film at a time when that was not popular and we were saying there was genocide in Tasmania — there were appalling things happening and there was the collusion of the scientific community in the nineteenth century in these acts in terms of classifying humans and racial anthropology. Unfortunately, we ran straight into the emerging land rights movement in Tasmania. Partly [it was] our fault — we should have perhaps stressed the point that the title, *The Last Tasmanian*, was really a literary allusion; it was like James Fenimore Cooper's *The Last of the Mohicans*. It didn't mean *the last person* — it meant something much more poetic.

I still feel that the terrible thing about Tasmania is that symbolically such a huge amount of human experience has been lost, and was lost in the nineteenth century, and might still be lost in Arnhem Land or in the Kimberley. It hasn't finished. All these languages are going.

Maybe now [1998] is to early to assess the film, *The Last Tasmanian*. Maybe in thirty years time. I certainly regret being put there ... At one stage, at the Sydney Opera House, they said, 'This is a racist film because it denies land rights to Aborigines'. That's not true.

Such were the debates in the early 1970s. And it had all taken off when Rhys, Jim Bowler and the great John Mulvaney had found the bones in Lake Mungo, New South Wales. Legend has it that they were so bewildered by the discovery of ancient human remains out there that Professor Mulvaney simply put them in a suitcase and just carried them back to Canberra and the Australian National University, in a manner that no one would dream of doing today. It also meant that the provenance of the skeleton was partially lost. How could you tell about the age of the immediate site if you'd removed the main prize? The bones are now being returned to where they were found.

Mungo 'Man', and Woman, as she turned out to be, were tens of thousands of years old, at least 30,000. This was sensational. No longer could we talk of new, recent arrivals, not around even as far back as the pyramids in Egypt, 5000 years ago. Instead we were scratching to work out how those ancient folk made it all the way to Australia from Africa without sprinting. And had they come in two phases, as the body shapes of those found later seemed to be either robust or slim (gracile).

And what about that *terra nullius* idea, that the Aborigines had just lived off the land without manipulating it and so altering it as European people had done? Suddenly archaeologists were finding fish traps, recognising 'fire-stick farming' and plenty of other clever ways that land had indeed been changed. Which led to the sometimes embarrassing question about who killed the megafauna? If people were here as far back as 60,000 years ago then they were without an alibi, though not proven culpable even now. Since that time, there have been tip-toeing public discussions on whether or not it was humans, Aborigines plus weather, or environmental factors alone. Tim Flannery says it was people; Mike Archer is more diplomatic. Every year the story changes as more evidence is found. Now, at least in the northern hemisphere, climate change is being blamed as a result of work by Alan Cooper and his ancient DNA team from the University of Adelaide.

It is a magnificent story of science changing our entire understanding of a group of human beings and, as a result, even changing our laws: Wik and Mabo stemmed as law on land rights directly from the newly discovered evidence. Yet the scientific story has become subdued. Partially this is because of Aboriginal peoples' demands that the remains of their ancestors be returned to them for burial. This despite the obvious huge separation of those fossilised individuals from modern people. I was saddened on visiting the American Museum of Natural History in New York years ago where the entire span of human evolution was on display from Lucy and her hominid relatives right through to the present, only to see an empty space where the Antipodean samples should have been because permission had been denied. There was not even a model of a skull or bones. Several archaeologists told me directly, though quietly, that they were giving up their work because the protocols to satisfy traditional owners were simply too vexing.

But amid the fraught politics about 'the intervention' in the Northern Territory, poverty and ill health, violence at home and outside, there is still a colossal amount of research outstanding. There is the immense amount of natural history known by the many communities across the land and how it should be recorded before it's too late. Harry Butler in Western Australia and many others have sung the praises of local, natural knowledge to me, but how much work is being done to write it down? Aboriginal people are easily the best natural historians, Harry told me on *The Science Show*. Then there is disease; it often seems beyond treatment, but think back to Dr Kerin O'Dea's experiment twenty-five years ago, when she took a band of diabetic indigenous people back to their land and asked them to try a traditional diet again. Within days, from fresh food obtained and exercise in going to find it, the diabetic symptoms had gone. It was a striking result.

> There was a remarkable improvement in glucose tolerance. Of the ten diabetics, six were no longer diabetic. The remaining four were greatly improved. Their glucose tolerance was almost in the normal range.

I think there were three major components — and this is terribly important for everybody to bear in mind — for the treatment of diabetes. It's not just diet, but certainly that was an important aspect of it. The second important aspect was a greatly increased activity level because, just by definition, having to go out and get your own food every day is a fairly energy-intensive process. And the third important aspect was weight loss. All of the subjects were overweight at the beginning of the study and all of them lost weight consistently throughout the seven weeks.

The other very dramatic observation made was that they had very high blood fat levels — their triglycerides were very, very elevated, which is characteristic of this type of diabetes. These were completely normalised.

I think there are three main principles that we are trying to incorporate into our future preventive programs. That is, if we take the important aspects of this study — the low fat diet, the increased physical activity level and the weight loss — these three things need to be incorporated into any successful program for the treatment of diabetes.

Kerin O'Dea is now professor of Population Health and Nutrition at the University of South Australia. Her work continues to explore surprising links between class and health.

Then there is the cultural significance of what is left. Here I mean more than the preservation of languages and some of the stories. How much do we non-Aboriginals understand how the song-lines, for instance, work? I was moved profoundly by Professor Nigel Spivey's series on ABC Television called *How Art Made the World*. He is a classicist from Emanuel College in Cambridge (where Rhys Jones was schooled). After showing how abstract art went back thousands of years — the Venus of Willendorf is hardly realistic and she is over 25,000 years old — Spivey lauded film as one of the pinnacles of art, combining as it does storytelling, music, movement, and even singing and dancing. Where did this sophisticated evolution come from, he asked, and then, astonishingly, referred to one of David Attenborough's films on Australian Aboriginal singing. It may seem to be guttural chanting to some, but it combines those elements in the song-lines, as do films: stories, music, dance, painting and much more. Are the subtleties of examples like that appreciated widely? I think not.

And there are the cave paintings, thousands of them, throughout northern Queensland, where Paul Tacon has tried to keep up with the vast challenge — and in the Kimberley, where Peter Veth, with Aboriginal assistance and advice, is working through the Kimberley Rock Art Foundation. The range of dates involved is also huge. The Bradshaw paintings alone could go back 50,000 years and are incredibly sophisticated.

One story will illustrate the significance of this art, barely known as it is, but to a few. New Zealander Mike Morwood, latterly, before his death, was at the University of Wollongong. He examined some of the rock art in the Kimberley and said, quite logically, this must have come from somewhere else — there is so much of it and the standard is brilliant. Many ancient people must have brought it to Australia. He looked north across the straights, which once had been narrow, and decided to go to the Island of Flores. There he looked for caves and found one with deep detritus, as if people had been living there for centuries. Then he dug, and discovered the Hobbit.

This was published in the journal *Nature* and has rocked the academic world ever since. Were they coeval with modern people? After all, they lived until 18,000 years ago. Why were they small, at barely a metre high? Was this a normal human or diseased? The informed view is not diseased — they lived there for too long and there were many of them. Anatomically, the ANU anthropologists insist that Hobbit is quite distinct. Now more signs of hominid relatives are being found on other islands.

One of the most sensational discoveries was made by Dr Stephen Munro of the Australian National University in

2015, also published in *Nature*; it was of an illustrated shell, clearly marked in a symbolic way, dating back 500,000 years! This is more than 300,000 years *before* the appearance of modern humans! *Homo erectus* in Java? How could our 'primitive' ancestors be so sophisticated that they used symbolic art? And what did it mean? Could this tell us about diets back then and the importance of molluscs and other omega-oil-rich foods that could have sustained brain growth and health aspects for those living in the littoral zone?

These are immensely important questions and we are sitting smack in the middle of the region where all of this detective work is happening — or should be if we had the resources.

And here's a final illustration from a group of New Zealand archaeologists. The scientists were fossicking in the highlands of Papua New Guinea when they found evidence of cooking fires and long-term habitation. But this was up in the snow line, freezing in winter. And the dates, again, went back 50,000 years. Who were these people living up there so long ago, and why did they forsake the coast? Was it because of mosquitoes and other pests? And how did they persevere in the cold zone?

It is a set of mysteries enough to make Indiana Jones's quests look tame.

* * *

And now for something completely different: New Zealand.

We may be 'cousins' but New Zealand is in every other way utterly different from Australia. We sit on different geological plates. The terrain in Australia is ancient, whereas that of New Zealand is brand new — hence the high mountains. The animals and plants are quite different too and we like possums but New Zealanders hate them. People arrived in Australia more than 50,000 years ago, but in New Zealand just 900 years ago, so very recently. The Maoris, 10 per cent of the population, have well-established rights and roles, which are much more elevated than those of our indigenous people — Maoris even have a treaty! And, as for that killing of the megafauna, there is no doubt about it — given the recent arrival dates, humans clobbered those giant moas, birds that were up to 2 metres tall, leaving none at all in this modern age.

I arrived in New Zealand in 1976, in Auckland, to find that the Maoris had occupied Battery Point, a swank projection of the city, equivalent to Dover Heights, a swank slice of Sydney. I duly asked permission of a female elder and recorded an interview of what was going on.

Then I repaired to the university. I was struck by two things. First, most of the professors seemed to be remainders of a 1950s British film of 'life on campus' and had the full complement of pipe, tweed coat, avuncular manner and even Scottish accents. It was an enclave in the far south of a pommy time capsule.

The second surprise was how accessible everybody was, whatever their station. One of my first interviews was in zoology, with Dr Cath Tizzard. She was married to the minister for defence and she went on to become the mayor of Auckland and then governor general. It did not stop her dropping into my house in Sydney at Christmas, despite her high office.

The last person I met on that first day was the professor of law at the Auckland University. I had puzzled about why I was being steered in the direction of a lawyer, but went along with it to be polite. Often, if you know you're wasting your time, you do a lightning interview in about seven minutes, thank the professor and go. The interview is never used, but it gets you out of the building fast.

Margaret Wilson, by contrast, was lively, talkative and very well informed. She was also younger than the other, tweedy academics. After the chat I suggested a beer. She got up and limped across the room to get her coat. 'Twisted your ankle?' I asked, guilelessly. 'No, I had cancer when I was young and they had to cut it off.' Gulp.

In the bar I got drinks and then noticed her eye was skew-wiff. 'Have you got some speck in that eye?' I ventured. 'No, I'm blind in that one...'

Doing really well I thought! But she was impervious. Then I discovered she was a TV star and was on *Beauty and the Beast* regularly...and then turned out to be President of the New Zealand Labour Party and was instrumental in selecting David Lange as the next New Zealand prime minister. Not bad for a first visit to the Land of the Long Flat Vowel.

And the science was excellent. Even after the Rogernomics cuts years later, when economic rationalism was applied by the pitiless Roger Douglas to emulate the equally scorched earth style of Reaganomics, the country managed to preserve those centres of excellence I have kept returning to for decades. Not least to Dunedin. In fact, Vice Chancellor Professor Harlene Hayne tells me their psychology department (her subject) is in the top ten in the world. I know it is an embarrassing cliché to keep referring to 'world class' but, in the case of Dunedin, it is true. Medicine, biology, geology, and environment all flourish there. And it is the world capital of penguins and albatrosses.

New Zealand is also a reminder of the relationships in science that cross all borders. Most of those in Auckland, Christchurch and Dunedin (and no doubt on other campuses) have firm connections with research in Oz, the USA and elsewhere. It shows that, even in a country with four million people, there is a momentum in science that cannot easily be thwarted. Yes, there are protocols also connected to Maori-related research, but they can be dealt with effectively.

People sometimes ask why I try to travel so much (notwithstanding the lack of ABC air tickets) and the answer, as Norman Swan will affirm, is that 98 per cent of the world's R&D goes on outside Australia — and you need to go there to test its true extent and direction. And New Zealand, despite its distinct origins and character, is firmly within our precinct but also a window to the wider world.

Just like with our story of the original Australians, that of the New Zealanders has grown and grown. Where did those Polynesians come from? Well, genetic analysis at the University of Victoria in Wellington has shown it is a combination of Taiwanese ancestors with Melanesian ones, the latter mainly men and the former mainly women. In this way that distinctive physical presence of people all the way from Hawaii to New Zealand has been formed.

And unlike that somewhat gung-ho approach discussed in the first *Science Show* with Thor Heyerdahl of actually sailing rafts from South America, our more recent investigators have instead traced the movement of coconuts, chickens and even language and pottery to work out the movement of Pacific people over the centuries. What is clear (and Thor Heyerdahl was, fair enough, correct about this fact) was that the Polynesians were incredible sailors. Just look at the size of the Pacific Ocean and the scattering of islands over immense distances. Even to Easter Island. Having been to Rapa Nui (Easter Island) once myself, staying for a few days, I know how long the plane flight was and the idea of reaching that dot on a map after weeks in a dugout canoe is truly awesome.

It is hard to believe that Aboriginal Australians, Melanesians in PNG and Polynesians in New Zealand and the South Pacific can be so similar and yet so different. And the story is still so very incomplete. I have watched it emerge in barely one career — not even fifty years. And the latest joke: having done a spoof involving the Fossil Beer Can, purporting to show that white men came from Australia and travelled north, now I see published evidence in the journal *Science*, that Australian genes have been spotted in the 'Indian' (that is indigenous) people of North America.

What a human history!

15 Marine Science

All at sea?

Today, as I write, the United Nations has let Australia off the profound embarrassment of having the Great Barrier Reef put on the endangered list. This, despite our having lost 50 per cent of the coral coverage in recent years.

Our reputation for marine protection is mixed. We do well with marine protected areas but poorly on development, with ports and industry coming first in the eyes of some political leaders. So, continuing this theme of coping with a country perhaps too large, how do you look after a coastline 36,000 kilometres long with an additional coast of 24,000 kilometres when you add the islands?

One answer is: we are still learning. My own encounter with marine life, growing up in Vienna — which has a limited coastline (!) — was on a day in the early 1950s (I think it may have been 1953) when I played truant and wandered around the beautiful but damaged city wondering what to do next. I saw the Urania cinema by the banks of the Danube Canal and decided to buy a ticket. The film showing had a title that did not, at first, inspire: something to do with Hans and Lotte Hass under the Red Sea. But when I went in — unbelievable — here were two people filming under water surrounded by crimson or blue fish, amid waving corals, all sights unknown hitherto, just after World War II.

Observing under water was totally new, unless you were a pearl diver in Broome or had some contact with the pioneering Hans Hass taking photos under the surface from 1938 onwards. Filming started in the late 1940s and the public began to see what could have been another planet, from roughly the date I did, from the mid-1950s. So the kind of up-close science you need to look after a reef and protect other marine systems has been possible only for sixty-five years. You may now take it for granted but it is so new.

Before reflecting on that, here are some thoughts from legends in the field:

David Attenborough on prospects for the Great Barrier Reef:

> Everybody knows how wonderful the Barrier Reef is, but perhaps not everybody knows what a key it is to the whole life of the oceans. If you lose those reefs, you lose the nurseries where the young of so many fish grow up. And the disaster would be just unspeakable.

And this is Sylvia Earle, one of the legends of deep diving:

> **RW:** I did an interview at the Darwin Museum and an oceanographer there said we know more about the backside of the moon than the tropical waters just off the north of Australia.
>
> **Sylvia Earle:** Yes, 5 per cent is about the magnitude of what we've seen of the ocean let alone explored. More people have been to the top of Mount Everest — hundreds of people have been to the top of that mountain — but only two have been a mere 11 kilometres down, and this is our own backyard.

Ove Hoegh-Guldberg from the University of Queensland on the threats:

> A lot of organisms in the ocean form calcium carbonate skeletons. Corals do it and they build the Great Barrier Reef. That's the housing for over a million species. If we go to the Southern Ocean, there are these tiny plant-like creatures called coccolithophores that form calcium carbonate skeletons and drive a pump, essentially, where carbon dioxide is absorbed by the ocean. They too are under threat. If we go to industries that depend on crustaceans, crabs and lobsters, those too are seeing really interesting diseases popping up, basically because things are not calcifying anymore. And therein lies the problem, that this equilibrium that we've had on the planet for 20 million years is horribly disturbed.

Professor Terry Hughes, head of the ARC Centre of Excellence on reef studies:

> There's been a huge response to the issue of overfishing globally in the last few years. One big response has been the establishment of so-called no-take areas, sometimes called green zones. When you catch a fish and remove it from the water and throw it into the Esky, you've obviously depleted the stock of that fish species by one, but you're also depleting the ecological function of that fish and different fish have different, sometimes very important, ecological functions on reefs.
>
> No-take areas are relatively easy to establish in in-shore waters where they can be policed and monitored. It's much more difficult to establish them in the open ocean but we're beginning to see that happen now.

After those key observations, here is a reminder, as Lynn Margulis announced at a conference in Oxford in 1993: 'Look at the world map. This is not the Planet Earth; it is the Planet Water.' And she was right. About seven-tenths of our planet is sea, ocean or lake and river, and they connect with currents, upwellings, fish migration, often across gigantic distances.

Both the chemistry and sea levels vary, depending on where you are. I found the sea level variation hard to believe (isn't it all more or less flat?) until I was reminded of Earth's gravity being different across the globe and the pull of the moon uneven through the day.

So it is hard to limit study of your own region when outside effects are so powerful. Even with the Great Barrier Reef — it really is 'Great'. It's 2500 kilometres long and stretches way out to sea and *downwards*. How deep can corals go and still flourish? What is their recovery after a cyclone? Why do their little algae, which provide their nutrition, depart when it gets too hot? Does that mean death for the polyp?

And, while you are pondering the GBR problem, lo, here's a reminder that there is a reef almost as big on the other side of Australia, along the west coast; one that has blue whales wandering close by. Just when you think you have the system worked out, a sensation is revealed, as it was in 1982 — that a full moon combined with the correct ocean temperature could trigger a mass spawning, during which most but not all of the reef creatures (such as corals) release their reproductive cells. It's like a vast cloud of living seeds. Divers would see nature applying the 'safety in numbers' theory to offspring. If everybody did it at once then survival overall would be increased.

As for the threats to reefs everywhere there is clearly the problem of Three. We have the chemical run-off from farms; then there is overfishing; finally we have climate change heating the waters and causing water to become more acidic. One of these might be survivable. Three together is too much. Add the odd cyclone and you have the accumulation of damage no ecosystem can survive. Going back to 'who killed the megafauna' in Australia, one can imagine a climate stress, a prolonged drought or some such setting that made the diprotodonts and giant kangaroos weaker; then add people with spears and clubs, and it's all too much for wildlife. Effects are additive.

But there is good news, as we have broadcast in *The Science Show*. Terry Hughes was at the AAAS meeting when he was able to announce his first five-year study of the effects of set-aside marine protected areas. The fish population *doubled*. But that was not all. Fish do not recognise our human boundaries and so many of them leaked into seas where fishing was allowed, so those who make their livelihood from fishing benefited too. I have explained before (in reference to lobster fishers at the Abrolhos Islands off Western Australia) that many industries do react positively when you explain the science and what is reasonably achievable. I am reminded of the Mexican fishers who were told that their unintended by-catch of a turtle species — the immature migrating specimens — was threatening the creatures with extinction. So they immediately made efforts to throw back any young turtles caught, and the problem was (mostly) solved. All it took was a friendly discussion plus the facts.

Farmers, too, in Queensland, have made efforts to reduce run-off from their plantations. But there is a long way to go. It is a pity that many on the land or industry perceive warnings about marine peril as the unrestrained ravings of the politically incontinent. That is why one must stick to the science. It is in our ABC programs, in the *New Scientist, Cosmos, Australasian Science* and other publications. Beware the internet, unless you find a site mounted by those bodies or reputable scientific organisations. But the science is known and it says that taking action is in the farmers' and fishers' best interests — and the nation's.

One of the ways I cope with the onslaught of confusing information is through the Australian Science Media Centre, now ten years old. A declaration: I am deputy chairman. The centre responds quickly, within hours, rarely days, to breaking science stories, such as the UN on the Great Barrier Reef. It provides three or four experts giving commentary on the implications of news. It may be another tremor in Christchurch, a cyclone on Vanuatu, or the efficacy of a new drug preventing HIV infection. I can look up such commentaries and get a rapid idea of what the received wisdom might be. It is not mandatory to follow their line, but it's a very good start. There are several quick ways for any citizen to find such guidance. It saves unnecessary panic or spleen.

Denial is another matter. Even some of the most well-informed people, and here I am thinking of a former editor of the hallowed *The Economist* science pages, will dispute figures on ocean acidification or even suggest that the corals or shellfish will not be affected adversely. Such assertions are easy to make. As my intrepid science author nephew Ben Goldacre, author of *Bad Science*, keeps reminding everyone, you can always find a professor somewhere who will declare the evidence to be at fault, and have a contrarian idea to confuse the populace. The Hobbit is an underdeveloped human; HIV does not cause AIDS; wind turbines cause migraine. It tells you something about the great variety of human cussedness but one really needs to take good advice and only then make a decision about risk.

The point about risk is that you must make an assessment of uncertain prospects. You don't know that the plane will crash or that the storm will blow your tree onto the neighbour's house or that the week-old chicken in the fridge is loaded with *Salmonella*. But you assess how much you and your family or community needs to do to stay safe. The minute you bring in others you are forced, unless you have psychopathic tendencies, to make the risk factor other than 100 per cent or nought. Nuance is required. Climbing on an unfamiliar motorbike when pissed obviously has a risk factor over 70 per cent. Getting on a Qantas plane, given their record, will be 0.0000002 per cent — safe. I was struck by the comments of Sam Mostyn in 2015 when she spoke at the University of New South Wales on a public panel (broadcast on 2 May that year) about risk. She told us both actuaries with insurance companies and some on the board she's on (this time a bank) were agreed about climate change in general: it is a clear and obvious risk with possible dire consequences and we need to take precautionary action urgently.

If you live in Queensland and have any number of industries connected to tourism and the GBR then you are risking a $6- or $7-billion enterprise. Not wise, whatever quibbles might surface about acidification. And, on the other hand, there are many dodgy industries we are going through the mangle to preserve which are, frankly, awfully old-fashioned. My father, down the pit in Wales, was digging up coal in 1919. It killed him at the age of fifty-seven, his two brothers at forty-two. The industry is still filthy. Hundreds die in coalmines every year, 164,000 since 1700 in Britain alone. Is it really worth that risk? And what about growing sugar in Queensland? Could we grow something else?

Are there no more 21st-century enterprises the Lucky Country can dream up and invest in?

It is rather like Japan and its whaling. One keeps wondering *why*? I did an interview at the University of Auckland with a marine scientist who was taking biopsies of so-called 'permitted' whales and finding that many were rare species.

> **Scott Baker:** If we look around the southern hemisphere and other areas, we see pockets where [whales] haven't recovered — why recovery here and not there? A possible explanation is the admission of this massive illegal whaling by the Soviets during the period following World War II. The 47,000 humpbacks they killed [and] didn't report clearly would have taken a big hole out of regional or local populations.

And then there was Dr Jonica Newby's investigation of the actual papers published of this alleged scientific side to the Japanese whaling. After the panel, including Mike Archer from UNSW, discounted those which did not qualify (no peer review, no reputable journal), their finding was laughable were it not so serious.

Mike Archer: From that whole pile of [research] papers we've got a total of four papers that can be said to be peer reviewed, that have some relevance to developing or managing a whaling industry and that also would require lethal sampling of whales to get this information. Just four papers.
Nick Gales: So with the eighteen-year program and 6800 whales killed, on that basis it means 1700 whales killed for each one of those four papers. It wouldn't pass an ethics committee.
Jonica Newby: So, what's your verdict — science or smokescreen?
Mike Archer: It's more like a smokescreen than science.
Robyn Williams: Professor Mike Archer, University of NSW, Professor Pat Quilty, ex-chief scientist from the Australian Antarctic Division, and Dr Nick Gales, University of Tasmania, with Jonica Newby from *Catalyst* showing once and for all the complete worthlessness of the Japanese 'scientific' whaling program; 1700 dead whales for each paper. And now the federal government is promising to confront the whalers on the high seas.

The science part of Japanese whaling is, frankly, bollocks. It is insulting and pointless. When I ask my friends in marine science why that sophisticated nation, Japan, should bother with such mendacity I am told they see whales as a stalking horse for tuna. They could not bear to give up their pricey sushi. So the farce continues. Defend killing whales so that the world doesn't stop you hunting fish.

Fisheries, as I have already indicated, quoting Professor Daniel Pauly, have been decimated. Yes, 90 per cent of the fish of yesteryear have gone. Pauly said: 'We have declared war on fish using all the equipment of battle — from sonar to fast boats and dynamite — and we have won. We have beaten fish as we would an enemy. But hang on — they aren't the enemy. In South East Asia they are bottom-trawling, which means scraping the bottom of a sea that has barely any living thing left down there, and when the trawlers have finished there is certainly nothing left. The industry survives, furthermore, by using slave labour.

This last piece of appalling information came from Professor Amanda Vincent, who, like Pauly, is at the University of British Columbia in Vancouver. She is known as the Seahorse Lady around the world. Once, she had a study of these exquisite creatures going on in Sydney Harbour (Watsons Bay). You know what seahorses look like but there are two additional pieces of information you need to know to comprehend Amanda Vincent's triumph. First, it is that the males give birth. After the female seahorse is fertilised and the offspring have grown a bit, they are transferred to the male. He distends, then gives birth a few days later, to a shoal of little seahorses. The second point is that the Chinese use them for some kind of ridiculous medicine, and they may be used in aquaria. The upshot is that fishers in South East Asia, especially in the Philippines, having no fish to catch because the pirates have dynamited them all, resort to seahorses. These they sell for puny sums, but enough, just, to feed the family. And this is where Amanda and her troops come in. They fly to the fishing villages, sit on the ground listening (of course they speak the local language, and more), then they tell the locals about the science. Catch adult seahorses and sell them and you are thus killing fifty or more animals. Preserve the adults, especially the pregnant males, and you put them in a fine mesh enclosure in the water off the beach and wait until birth time. Then fifty to a hundred little seahorses swim off through the mesh, leaving the adults, which you can duly sell. Everyone wins.

The next thing is the neighbouring village, noting that yours has plenty of catch, comes to raid you, but you intervene, tell them about the science, and they follow suit. Welcome to the R&D of the future. Tomorrow's scientist, in this field anyway, needs to be multi-lingual, diplomatic, enterprising and cool. I have seen Amanda go to a Chinese medicine shop where (dead) seahorses are piled high, and she is as serene as Mother Teresa. I'd be tearing the walls down.

* * *

But I am already tearing the walls down about seagrasses. Seahorses live in them; dugongs, abundant off the Top End (and studied by Helene Marsh at James Cook University for forty years) eat them; and they grow along 92,500 square kilometres off the coast of Australia. Or, they used to. They are being killed by water pollution, development and will die if sea levels rise too quickly.

Why should you care? Because seagrasses absorb *forty times* as much carbon dioxide as tropical forests. Maybe more. About 150 million tonnes of carbon are stored in Australia's seagrasses. About 0.2 per cent of the world's oceans apparently still display seagrass beds and we need every square centimetre. (Science is full of stats like this.) They offer solutions as well as challenges. Preserving seagrasses is relatively easy. Just leave them alone. Another example is those holes in the ground that have been on fire, sometimes for years. Morewell, in Victoria was an egregious local example. The Hazelwood mine burned for 45 days causing major health problems. The fires in Indonesia in November, 2015, caused severe smog in Singapore, burned for weeks on the underlying peat and covered an area twelve times that of Greater Sydney or Melbourne. But my point about solutions is that these fires amount altogether to nearly 3 per cent of greenhouse gases, close to the total amount released by aircraft. Now, is it easier to put out fires in holes or to close the air travel industry?

In 2012 I interviewed a young woman from Kuwait doing her PhD on these seagrasses. It was at the AAAS meeting in Vancouver and she told me she was keen to convince the Gulf States that their fisheries would collapse if they did not protect offshore grasses. The grasses are not only precious nurseries for fish, they also absorb 40–100 times the amount of carbon dioxide as do tropical rainforests! On the Monday after the broadcast I received an email from a fourteen-year-old lad living in Kuwait. He told me he listened to *The Science Show*, was galvanised by the seagrass story and had asked his mates to join him in a schoolboy/girl study to take things further.

It's what makes doing these programs worthwhile. We are no longer just talking to Oz. Every week we're in touch with the world! And, as far as the ocean's concerned, it may make all the difference.

These are some of the stories we broadcast about Planet Water. They show what outrageous lengths some villains will go to for an extra buck. But they also demonstrate the ingenuity of the scientist — often combined with citizen science — to solve a problem. This could be a force to be reckoned with. If we have time.

16 Germs!
The ultimate act of sharing

As 2015 was the Year of Soil, allow me to offer one of my favourite factoids to celebrate; one I have used more than once in *The Science Show*.

Cup your hands and fill them with damp earth. If you were to count the microorganisms you are now holding so easily it would come to more than the entire number of human beings who have ever lived on Earth — more than eight billion. Now look at your garden and think about it! How many are in there?

All those viruses, bacteria, slime moulds, fungi — be pleased they're around, otherwise we'd be in trouble. And, by the way, *you* consist of a trillion cells but also play host to ten times as many germs: TEN TRILLION! You are a walking, talking mega-infection. Again, be pleased. We could not live without them.

We are fairly pristine before we're born but pick up plenty of microorganisms on the way out, travelling through our mother's passage. It seems that Caesarean sections do deprive baby of some of these essential symbionts. Gradually human contact adds to the complement of germs we accumulate and we end up, unless we're too fastidious, with the number of bacteria we need to stay healthy and functional. This is one reason I have always felt that cunnilingus is a good health measure.

The newspapers are, these days, packed with stories about the microbiome, and most of them are fair enough. It is probably thirty years since I first was offered what seemed to be an unlikely story involving the juxtaposition of back pain with bacterial infection. Then, the researcher trotted out another list of ailments he linked in the same way to germs — some good, some bad. That's when my interest faded — claims can be too large and diminish the impact of focussed thought. But that thought has not gone away.

Now the science has exploded. In early 2015 Norman Swan broadcast a *Health Report* based on a huge meta-study (putting lots of published works together) showing that several mental illnesses are now associated with diet and the ensuing bacterial population in the gut. The study came from Deakin University, was peer reviewed and published in a reliable journal, but even Norman said, on air, that he's had his doubts before but was now simply astonished. Germs and psychosis, germs and depression, germs and schizophrenia — what next?

Faecal transplants? We have featured them in *The Health Report*, *The Science Show* and *Catalyst* — oh, and *Ockham's Razor* too, in case you thought we had neglected the issue. I first came across the story on NPR (National Public Radio) in the US. Having just arrived in New York, I was listening to *Radiolab* on my headphones. They told the story of the Deep South of the US where agricultural production had been hopelessly low and the workforce seemed almost somnolent. Was it that the African-American inhabitants were now lazy, having escaped their slave status? No, according to the program it was because they were far too prone to illness. Their immune systems had crashed. The answer? It seemed that normal infection from pests such as hookworm, which stimulated the immune system, were missing in southern USA. When hookworm was reintroduced, the population thrived. This was the thrust of the program — which also featured an Englishman who had similarly become overwhelmed by lassitude and debilitating ailments preventing him from earning a living. He read about the hookworms and decided to fly to parts of Africa where they were common in mud. He walked extensively in this unappealing swamp and, when sure he'd been infected, he sat back to see the result.

He recovered, and was back to his old lively self. His immune system had been rebooted by the presence of the otherwise harmless little parasites. Then he had a thought: why not sell, even export, this faecal salvation? This he has done, in dried form, including hookworms, carried by FedEx all over the world.

Now faecal samples can be offered in a capsule, no more repellent than the average pill. And the idea that we can replenish our internal biome after, say, taking too much of a hit from antibiotics, may be a good idea. Some diseases that could kill were the first to invite the faecal solution; now the measures are being considered for wider purposes. Then there are the antibiotics given to cattle and chickens to make them grow faster — what are they doing to our precious bugs? An outrageous problem when you consider that they are not needed. At all.

* * *

You can blame Lynn Margulis. This unbelievable woman, once married to Carl Sagan, whose son she wrote with (latterly) and half her ideas were brilliant. She had rows with the editors of journals who discounted her unblushing propositions and she fought the scientific establishment to the very end. Her main thought was that there is as much cooperation in nature as there is competition. She cited the compelling example of all those microorganisms that lived together in a closeness that eventually led to enclosure of one by another. The bugs joined forces; they merged.

The mitochondria we have in our cells were once free-living bacteria of some sort. They invaded other tiny creatures and became incorporated because they were useful. Mitochondria are the energy packs of our bodies and those of countless other animals. They maintain their own special DNA systems, distinct from ours, but smaller than they once were. We inherent mitochondria only from our mothers — sperm have a few but they are not absorbed into the egg during fertilisation. So we are full of those former beasties. The same can be said about chloroplasts in most plants, once also separate organisms. This is another enormously successful alliance.

Look at corals, with their tiny algae, providing carbohydrates to their hosts and needing to be in reach of light to keep their photosynthesis going. Lichens? They're a combination of fungus and other algae. And so went Margulis's story, showing combinations count and maybe multicellular organisms came from mechanisms allowing the cells to stick together without having to go their own separate ways. But that did not happen for two billion years after the origins of life in the deep vents at the bottom of oceans. For all that time, the Archaea bacteria and viruses (ancient microorganisms) just kept going on their own. Some of them, when enlarged, look spectacular, but the planet was essentially ticking over, as far as we're concerned, until one and a half billion years ago when the multi-cellular experiment took off.

And as we grew and developed all those organs and fluids we maintained those little guests. We had to. They were everywhere, inescapable. So we'd better come to an agreement with them and help each other. We gave them a home and nutrition. They provided vitamins and stopped nasty bacteria taking over — when we're doing well. If we do have an invasion of nasty bacteria, our own often fight to keep them out. But the newcomers only represent fresh invaders we're not used to. Add a few hundred years and, you never know, they could stay as yet more harmless lodgers. It is ultimately not a good idea to kill your host, or you have to invent all sorts of complicated mechanisms — like hitching a ride on sneezes — to carry on your bugged life cycle. Adaptation to become benign is an advantage both ways.

As for those antibiotics given to animals we eat, we ran a story in 2014 that was a follow-up to an article in *Scientific American*. We could do so because we had, for once, an intern from the University of South Australia who could look into it for us. The story suggested that, at least in America, the drugs were given mindlessly, because a tradition was being maintained. The point was: they don't work. Young farm animals may grow more quickly, but overall the weight ends up the same. Yet vast tonnages of antibiotics are fed to them, a large percentage of the American total.

[Reporter] Isabella Pittaway: America, 1955: a crowd has gathered at a hotel to watch [animal] feed salesmen be weighed. The men have been competing for four months to see who could gain the most weight eating food laced with the antibiotics they sell for livestock. American pharmaceutical company Pfizer sponsored the competition.

It's long been a practice in the agricultural industry to feed animals antibiotics to make them grow bigger faster. In fact, we can trace it back to the 1940s when American biochemist Thomas H. Jukes first came across the discovery through an experiment with chickens. Since then, antibiotic use in animals has increased, and it's estimated 80 per cent of antibiotic sales in the US is for livestock. In Australia only a third of all antibiotics consumed are for humans, while 350 tonnes of the drugs go into stockfeed annually. Have we squandered these precious lifesaving drugs by using them for growth promotion in animals? And are they even necessary? Antibiotic resistance is a global public health issue, and the misuse of antibiotics in animals is one of many contributing factors.

Richard Conniff (USA science writer): There is very little serious scientific study where [scientists] look at the long-term effects of antibiotics on the animals and demonstrate that they are either economically beneficial or beneficial health-wise. We're just going on the drug industry's word that this is the way it should be and this is how it works.

There was a study done by Johns Hopkins University a few years ago in which they took results from a large producer of chicken in this country (USA). They could find no economic benefit from using antibiotics; that basically the cost of the antibiotics was equal to the gain in weight the chickens had as a result. There have been other studies by the US Department of Agriculture, which have found similar results in other animals.

IP: So why are producers still putting antibiotics into their animal feeds if there are economic losses and there's not that much of a change in growth?

RC: Agriculture is a pretty conservative industry. It has been an article of faith for sixty years now that antibiotics work to put on weight on animals, to make them get to the market weight faster. The other thing is that using antibiotics allows them to crowd livestock into industrial-style manufacturing facilities, and they've become dependent on that style of production. So the idea of turning away from antibiotics is really frightening to the livestock industry. It requires a whole new way of thinking.

IP: A recent review by the Food and Drug Administration found that of the thirty types of antibiotics used in animal feed, twenty-six had never met the original safety criteria established in 1973. In addition, eighteen of the drugs were found to pose a high risk of exposing humans to antibiotic-resistant bacteria, and are allowed to stay on the market. What sort of data and surveillance is available on antibiotic usage in the US?

RC: Almost none. Even the Food and Drug Administration does not have reliable data on what antibiotics are being given to which animals and where. We have much better information about human use of antibiotics. But the livestock industry has been incredibly secretive about its use, and the pharmaceutical companies that profit from this business have also been highly secretive.

Australia, it seems, is more responsible, but still, consumption on this scale, in a way that risks a permanent overload of drugs followed by the depletion of our microorganisms, will surely turn out to be stupid and costly. What are we waiting for?

A postscript to that story, a posting from the Australian Science Media Centre: 'For immediate release: Worldwide antibiotics use in the livestock industry (PNAS) An international study has estimated that around 63,151 tons of antimicrobials were used for livestock production in 2010 by 228 countries worldwide, including Australia, with this amount predicted to increase by 67 per cent by 2030.' PNAS, by the way stands for proceedings of the National Academy of Science, in America. And the authors say their work indicates where hot spots of resistance my occur and where, therefore, extra attention should be given to amelioration. An increase of 67 per cent! This is insane.

As for all those allergies that are multiplying, it turns out that our upbringing should be more relaxed. You need not exactly roll baby in the dirt on the floor but do allow some leeway because it is invariably harmless. Like the African-Americans with hookworms, the added germs can help you build up resistance. Real infections involving disease are a different matter and hand washing remains a good move, especially when near hospitals or crowded areas like train stations or on escalators. Other than that, go grubby. But I must say I find the universal habit of handshaking incomprehensible. When I was on chemo, with a depleted immune system, I tried to replace the handshake with an ET-type little finger touch. But even when I explained my prudence, people still got up close. The contradictions are astounding.

Still, this invisible life has an excellent educational role. Just as Leeuwenhoek, who first examined his 'animalcules' when peering down his newly constructed microscope in Holland was mesmerised by their antics, so pupils in schools need to go back to this tradition. No need to go straight to a screen — leave the devices alone! Just go to the pond, or your mouth, get a specimen and see it teem with life. Much of that life is you.

Here is what Antonie van Leeuwenhoek wrote on what he saw, in a letter dated 1674:

> I took up a little glass phial [of lake water]; and examining this water the next day, I found floating therein diverse earthy particles, and some green streaks, spirally wound serpent-wise, and orderly arranged [identified as the common green alga *Spirogyra*: the earliest recorded observation of this organism], after the manner of the copper or tin worms, which distillers use to cool their liquors as they distil over. The whole circumference of each of these streaks was about the thickness of a hair of one's head. Other particles had but the beginning of the foresaid streak; but all consisted of very small green globules joined together; and there were very many small green globules as well. Among these there were, besides, very many little animalcules, whereof some were roundish, while others, a bit bigger, consisted on an oval. On these last I saw two little legs near the head, and two little fins at the hindmost end of the body. Others were somewhat longer than an oval, and these were very slow a-moving, and few in number. These animalcules had diverse colours, some being whitish and transparent; others with green and very glittering little scales; others again were green in the middle, and before and behind white; others yet were ashen grey. And the motion of most of these animalcules in the water was so swift, and so various upwards, downwards, and round about, that 'twas wonderful to see: and I judge that some of these little creatures were above a thousand times smaller than the smallest ones I have ever yet seen, upon the rind of cheese, in wheaten flour, mould, and the like.

Leeuwenhoek: the man who first showed the ubiquity of our very small guests and neighbours as he looked down his lenses. We can never escape them, not that we'd want to. Howard Hughes did try and lost his mind in the process. What a waste, as is the obsession now in keeping our children too clean and protected from too many foods, allowing them to become subject to allergies later on. A relaxed exposure down the microscope when very young may change this. The world to come, with its recycled water and drugs made of shit, will need a more flexible mind, a more mature way of dealing with what kids call the 'Yuck Factor'.

A final thought: I have always been amused by the homeopathic proposal that water has a kind of memory-preserving quality, that active ingredient remaining somehow in solution to 'cure' the patient. But just think — the water has been through a million toilets and, who knows, another million bowels. Perhaps it is the 'memory' of all those gut bacilli that is the secret.

17 The War on Science

Who is winning?

This question has come up in nearly every chapter: are scientists and our programs too consumed with warnings and worry, and not enough with solutions and a hope of progress? An example of the latter approach is Matt Ridley in his book *The Rational Optimist*, featured at length on *The Science Show* when it was first published.

Matt says we have gradually improved life through an unfettered market and individual freedoms so that now we have painkillers, longer life spans, more trees, productive agriculture using fewer chemicals, and a world population getting steadily richer and with less hunger. On the other side, as these chapters have indicated and we've heard from *The Science Show* No.1 onwards, there have been animal extinctions, nuclear arms races, dying oceans, extreme weather and sundry other potential miseries. What do we make of the divide?

My own feeling is that we have here a home called Earth, which happens to have three times the number of people it can comfortably accommodate. Imagine that multiplied population in your house or flat, and the rest follows. If I have twelve people in my little house, Termite Towers, the loo would break in no time, the kitchen would blow up and the noise become so intrusive we'd all go bonkers. And that's on the first day.

But you could say, à la Ridley, that there are plenty of open spaces everywhere and, if you did it cleverly, as the Arabs have done in the deserts surrounding Dubai (often using near slave labour), people could live all over the place. The world seems crowded, but the entire seven billion population of Earth could stand shoulder to shoulder in Tasmania and still have space to spare. Let the entrepreneurs and innovators get on with it and we shall be fine. Progress will, as ever, be unpredictable, but it will come.

Bjørn Lomborg applies a similar argument to Ridley's optimistic line. He says we must choose the life-saving problems we can solve *now*, ones much more affordable than climate change measures that may prove to have worked by 2050 or later but on such a small scale as to be nearly worthless. Bjørn has been on *The Science Show* many times, I enjoy his company — but he always says the same thing. Kerosene lamps and cookers in slum huts is one of his favourites, in that we could provide cheap, clean alternatives today and make a gigantic difference to wellbeing. The emancipated poor could then break free, work, and make further solutions in other areas more affordable. It was the line taken in the film *The Great Global Warming Swindle.* Thin children with big dark eyes were shown being thoughtful about middle class Western greenies wanting to deprive Africans and Asians and some South Americans of the rich plunder and fun games we privileged boomers had enjoyed in the past fifty years. What about *our* turn, those eyes seemed to ask. The commentary certainly did.

But what are Ridley and Lomborg, and our former ABC chairman Maurice Newman, asking us to do? Essentially, it's 'carry on as usual'. Let the market flourish and we shall find solutions as we always have. Blunder on through! Lomborg says invest in green technology. Well, some are.

Matt Ridley has now joined Nigel Lawson's Global Warming Foundation, a lobby group that relentlessly disputes climate change science. Ridley claims to be a 'luke warmist', as does Lomborg, but, unlike most of the scientists I talk to, all three have the same message, year after year, just like the politicians we keep hearing ('stop the boats, drop the tax'). Sometimes I think I am reading a newspaper from two (or five) years ago — here comes that bit again. It is accompanied by single-word insults: 'alarmist'; 'greenie'; or simply 'leftist'. Such repetitive writing, often in whole slabs of newsprint, has the same effect as racist abuse once had: it condemns a whole section of society — the intelligentsia, academics, public broadcasters, scientists — as some kind of coordinated movement, acting against the nation's real interests.

I doubt that a majority of Australians take these tribal attacks seriously, not least because they come almost uniformly from one (smallish) section of the (wealthy) population. But the effect is undeniable. This continuing onslaught has shaken public confidence in scientific culture, if not consciously, then at least in support from young people in their studies and the constituencies backing leaders who want to prepare for a rocky future. It has become a War on Science. In March 2015 this was the very headline on the cover of that otherwise restrained magazine *National Geographic*. It was also the title of a BBC film fronted by Sir Paul Nurse when he left America to become the president of The Royal Society of London. Both cited attacks on climate science, evolutionary theory and vaccinations as main worries.

Both Nat Geo and Sir Paul also cited green groups and their unwavering resistance to GM (genetically modified) crops. Such antediluvian attitudes to the research (not the politics, which is a different matter) does nobody any good, not least environmentally concerned activists who will pursue one line, and here Lomborg is right, however tangential it is to the main story. It is time that green groups got real, got coordinated and, as Greenpeace Australia has managed to do in recent years, got their priorities worked out. Politics remains the 'art of the possible'; mere shouting does not. Greens could also take the science more seriously and combine with business to effect compromise and results, not merely stand-off and purity. Brendan May, who chairs Robertsbridge, a British outfit that works with corporations internationally to make them ditch 'greenwash' and face environmental realities, has said as much on *The Science Show* many times.

Matt Ridley is right about many improvements in how we live. Most of them came from concerned people presenting honest research and saying change must happen. Lead in petrol and paint, smoking, ozone, asbestos, mercury, obesity — we mounted broadcasts on each issue, quickly followed by opposition from the same quarters eventually beaten by the sheer weight of evidence. Who now will defend asbestos, tobacco or lead in petrol? They did so not all that long ago. There comes a point when the public appreciates something must happen.

There is a case for limiting growth. This will mean gradually phasing out nineteenth-century industries like digging coal for burning. Coal should be the basis for a chemical industry (those carbon fibres, that grapheme technology, the methanol enterprise being researched at Monash, new organics), which is worth so much more and therefore not being consumed in power stations in America where 60,000 children get mercury in their tissues as a result. Imagine all those new batteries, solar devices, water cleaners, handy robots, mind-controlled hands and feet for paraplegics. There would be new products demanding less expensive content in decentralised industries, redefining growth instead of making it like it always has been — a relentlessly enlarging grey blob of activity, any activity.

Green, in the sense of environmentally sound in sensible ways, can also be efficient. The last series of Danish TV program *Borgen* showed the difference between revving the crowds with empty rhetoric and instead offering them a fully costed and tested alternative to pricey poles and wires, filthy fuels that kill and products people don't need as opposed to those they might really need.

The War on Science is being fought in many ways. Other than the public noise and propaganda, there is the constant use of complaint to tie up organisations that produce evidence about risk.

Michael Halpern referred to it in an interview from the AAAS, on *The Science Show* in March 2015. His study gave egregious figures for the calls and demands made of climate-connected organisations in the USA and Australia, showing how the law often requires us to answer objections — or even demands for information — that take minutes to send and often days to answer. The broadcaster, government department or university office becomes paralysed and eventually even fearful of every inquiry.

> **Michael Halpern:** When scientists decide to pursue contentious lines of research, whether it be climate change or chemical safety or food security, a lot of the time they come under an enormous amount of scrutiny, and some of that scrutiny is deserved. We should understand their research methods and how they do their work and whether or not they have funding from entities that would compromise the work that they do. But a lot of the time there are people who harass scientists because they don't like their research results. This type of harassment can come in many different venues, and we've seen an increase in the use of open-records laws, which are designed to figure out how government works, to try to invade the privacy of researchers and to try to distract them from doing the work that they do.
>
> There are thousands and thousands of requests submitted around the world every day. We know that most of these are good requests, submitted in good faith to learn more about how governments make decisions and whether or not there are inappropriate influences on those decisions. What people have started to do is use these open-records laws to go after university scientists who are studying very complex issues

and are often working with colleagues around the country and around the world.

Even if scientists work in a place where they feel protected, where their university understands the work they do, their work can be caught up in open-records requests in other places. So if I'm working at the University of Canberra and someone else is working at the University of Texas, if the Texas open-records law is very expansive, the Canberra researchers' documents can be caught up in that sort of thing. This is really a global problem that the scientific community is starting to pay attention to and universities have to deal with as well.

The use of open-records laws to harass climate scientists transcends international boundaries. In Australia one organisation submitted more than 750 open-records requests to the Federal government's Department of Climate Change, accounting for more than 95 per cent of all requests the department received. Staff members would have needed nearly forty hours to process each request. These repeated and excessive requests can be compared to denial of service attacks perpetrated on internet users. That is, they essentially flood an office with requests, greatly slowing down other work.

My own rule is rarely, if ever, give a platform to a lobby group (green or brown), and base reports wherever possible on well-regarded scientific work preferably as published in a leading journal. Succumb to the lobbyists and, as I've said before, you get both shameless repetition and wilful ignoring of factual answers we've broadcast many times before.

This kind of byplay is daily and unrelenting. I and other colleagues have no personal assistant to handle the deluge. We just start work at 6 a.m. instead of 9 a.m. to deal with it.

Should we be made of sterner stuff? Well, we are. But our organisations, having been under assault for many years, have now bowed to pressure from Canberra and certain media to be ultra-cautious about what we say on some of these topics. You may laugh, having heard live shows mouth strong language about radical ideas. Maybe — but in the unending stream of 24/7 multi-network broadcasting, tweeting and on-line blogging, you are bound to find a loose comment, a statement needing qualification. Then you are up for it. But times may be changing; I sense both the ABC, SBS and others are getting fed up with it and realise that responding makes no difference.

The real test is whether our programs are outside the flow of reasonable discourse in Australian society. I find them, by that standard, to be rather tame.

Some of the above was ever thus; much is new and deeply disturbing. Government is increasingly powerless to effect necessary and urgent change and the media make it harder by conjuring an atmosphere of antagonism and distrust. I believe the goodwill of the vast majority of the Australian public is solid and admirable, as was demonstrated after the Lindt Café siege in Sydney's Martin Place and the arson perpetrated in my own suburb of Rozelle when a convenience store owner allegedly burned his shop down, and (also allegedly) killed a mother, her baby and another neighbour as a result. People in their thousands came out carrying flowers and toys and wrote that this will not make us wilt or hide; we are together and we want decency and hope. They also, by implication, wanted the scaffolding of a civil society to offer opportunity and fairness.

It all makes sense if you go back to those arguments about risk. You may avow any wild idea you like if representing only yourself, like some despot in *Game of Thrones*, but once you represent family, neighbours, friends, then you must calculate the risk. Forget those puerile absolutes. Nothing is 100 per cent or zero per cent in nature. Try nuance. Climate may be a problem. Face it and do the numbers.

This was Malcolm Turnbull on stage at UNSW shortly

before becoming PM in a discussion on 'Risky Business' which I chaired and then broadcast on *The Science Show*.

> We cannot run away from risk or uncertainty. We cannot be afeard of it. We have to embrace volatility; we have to ensure that we are always as agile and flexible as we can [be]. Business people say, 'We just want certainty', and I always say, 'Well, you've come to the wrong place'. And then they say, 'What, Canberra?' And I say, 'No, Planet Earth. Wrong planet, sorry.' We are living in an inherently uncertain environment. But I think the key is being able to manage that, and that means that we shouldn't imagine that the way we did things yesterday is the way we should do them tomorrow. We've got to be prepared to review our approach to business, to policies and politics, for example, all the time, on the basis of the evidence and experience.
>
> Embracing volatility and flexibility is critically important because that is how you manage risk.
>
> Fear is a very big emotion. Human beings are not entirely rational. I think it is absolutely critical — it's a critical responsibility of governments — not to promote fear. And in fact particularly in the face of terrorism, which is a real threat (we are not kidding ourselves about that) but we've got to recognise that what the terrorist wants to do [is] frighten us. So fear is the outcome the terrorist wants. Terrorism is the propaganda of the deed. They want to frighten us and make us go about our lives in a very different way. So it's very important to be strong, to be united and not to be frightened. And the truly inspiring way this country and this city of Sydney in particular pulled together after the Lindt Café siege was remarkable. This is a city that came together with a practical solidarity founded in love, not in fear. That was exactly the right response; that is exactly the opposite of what the terrorists want.

On becoming prime minister Malcolm Turnbull elaborated this theme. At the same time as warning against the fear of uncertainty and of bullies (such as terrorists) he invites us to embrace innovation and forward looking. His warnings against the propaganda merchants fuelling that war against science have been less forthright so far, but I'm sure they are not far away. The stakes are too high.

18 Here Comes the A-team
It's bigger than you think

I have presented *The Science Show* for forty years now and I often wonder whether my acute awareness of the assaults discussed in the last chapter are of no account to the typical Australian citizen. A friend of mine, head of news at SBS TV, is convinced that all the attacks on his network and mine are of no interest to anyone beyond those in the industry, and he may be right. Which also raises the question: if I weren't immersed in science programs, would I have noticed much about environmental destruction or even reports on it in the media? I could well have glanced at the cityscape and the rural scene and thought not much has changed. No obvious signs of mayhem beyond gridlock and bare hillsides.

I could also ask a similar question about support for science. The casual onlooker might not often go to campuses or the CSIRO, but if he or she did, the premises would look vast and well appointed, plenty of cranes and new buildings, from the 'cheese grater' in Adelaide, where medical R&D is now underway, to UNSW where new shiny buildings abound and even the University of Sydney where sandstone is being matched by the glass and steel of new labs, as it is at the University of Queensland. Who would know that science is a dodgy prospect for young people? As with green stuff, your eyes skim the headlines, barely penetrating the articles. And you do vaguely remember that there's a Future Fund in the federal budget worth $20 billion, yielding a billion a year for medical research when it is established in five years or so. We hope. So it's up and down, but life goes on. Is it a matter of occasional austerity or fatal neglect? It seems like every day I see headlines in the newspapers, like this one, next to a picture of Catherine Livingstone, president of the Australian Business Council: 'The single biggest human capital challenge for the country is underperformance in science, technology, engineering and maths skills.'

Let me be plain. I am annoyed by Australia's lack of foresight, neglect of young talent and smug apathy about our scientific prospects, mainly because I am so much aware of our superb standards in so many areas. I have my heroes and know that the older ones (I kick off below with the oldest) can easily be matched by the young ones if only they're given a chance — half the chance that all the money merchants, hassling spivs and other untouchable (it seems) partners in Dodgy Brothers Inc. get away with.

Earlier on, I gave a brief list of the leading edges of research in various parts of Australia. Here, now, is another list of heroes and their institutions. It is not, of course, comprehensive. There are many campuses I have not been to in years: Charles Sturt, Darwin, James Cook, University of New England, Griffith, and so on. And there are lots of high achievers I've also not encountered but hope to. Take this more as a haphazard exercise of putting pins on a very large map. And we begin in Canberra with:

Dr Jim Peacock, CSIRO

Jim got very cross when applying for his degree course. He wanted economics, but that was full up or unavailable for other reasons. So they stuck him in the sissy stream: biology. No real men did pretty plants and furry animals back then, so Jim was miffed (not a pretty sight), that is until the magic of *modern* life science struck home — those genes, that maths, the chance to breed new combinations of characteristics. And then, think about it, from plants comes most of our food. It is risky as we depend on such a small range of cereals and root veggies, all with like genomes and all vulnerable to wipe-out if a new plague lands as it did on potatoes in Ireland in the 1840s. Today citrus and bananas are frighteningly open to catastrophe.

Jim achieved all the top honours: Chief Scientist of Australia, president of our Academy of Science, the PM's Science prize — this last award for showing how flowering is controlled in most angiosperms. His aim is to help feed the world as our population soars to at least nine billion. He is relaxed about GM and points out how genetically engineered cotton has done so well, eliminating the use of much water and many chemicals.

But Jim is well aware that investment is down and the paradox this implies. Plants mean not only food but medicines, rubber, timber and shade — in other words 'wealth'. That is why Kew Gardens, now over 250 years old and a centre for excellent R&D (including the use of seed banks), called its recent BBC radio series and book *Roots To Riches*. Vast riches. So why, as the book came out, did the UK government slash Kew's funding, leading to the firing of many good scientists? Myopia? Stupidity? Here we go again. It's not only in Oz where daft deeds are done.

But Jim Peacock has, throughout, been a visionary. His Discovery Centre at CSIRO in Canberra is a testament to that. And his achievements, and those of his colleagues, speak for themselves but should have wider recognition.

And before leaving the botanists: Professor Alan Kerr from Adelaide was the first local researcher to receive a PM's prize (then called the Australia Prize) for work on snipping DNA and making GM possible.

Harry Messel, Physics, University of Sydney
'I want your money,' the unstoppable Messel told Frank Packer, father of Kerry. 'What's in it for me?' snarled the captain of publishing.

'Nothing,' said Harry. 'OK, the cheque's in the post,' replied Packer. So goes the legend. And it was. The cheque was a huge one.

Harry Messel: The foundation that I set up, the Nuclear Research Foundation, was the first foundation in the British Commonwealth, not just the first in Australia — we were number one off the bloody mark on this.

The University of Sydney had just finished its centenary [fundraising], which was a great flop. It only got £50,000 or some damned thing. The very first person I interviewed was Frank Packer of Consolidated Press. I up and grabbed Frank Packer sitting across the table from me looking tough as hell but friendly-like on this. And I went and gave him the great spiel and I said: 'You know, Frank' — he wasn't knighted in those days — 'we've got to train and retain our own young people and I want to set up a foundation which is going to provide the funds in order to do this.' And I carried on for a while and he says: 'How much is this going to cost me?' And I said, I think, it was £2,000 a year. And he said: 'And what am I going to get for this, Professor?' I said, 'Right, I can give you an absolute guarantee.' He says, 'A guarantee?' I said, 'Damned right, I can guarantee you'll get absolutely nothing.' And he said, 'Get out of here, the cheque will be in the mail.'

And there was the first member of the Nuclear Research Foundation: Frankie Packer, God bless him. What a fantastic thing.

The same [thing happened] in those days [mid-1950s] about the first computer. When I got it, everybody was telling me to stick to my slide-rule. Today everybody has computers; you have them in your bloody pocket; you have them around your ears.

When I mentioned the idea of an electronic computer everybody pooh-poohed me again — oh, why don't you stick to your old slide-rule et cetera; what do you want these newfangled things for? But at that time one of the scientists I had appointed was John Blatt from the University of Illinois. The University of Illinois happened to have ILLIAC [Illinois Automatic Computer, one of a series of supercomputers], the brother of MANIAC at Los Alamos, which had been used for the atomic bomb design.

So he got the computer as well! Harry Messel was the first, really, to exploit private funds. Through this he built a phenomenal Physics Department at the University of Sydney, a famous summer school for kids, and 'the blue book', the weighty science textbook nearly every pupil used for generations.

Purists were, at first, sniffy about the use of independent dosh for a traditionally government funded educational system. Now it is the essential add-on. And that drive to get things done? — Messel was another pioneer. No wonder the crocodiles stopped grinning when he went north to study them by developing new tracking systems.

Harry died in July 2015 aged ninety-three. He was stoppable after all.

Among his appointments was my late friend Robert Hanbury Brown, professor of Astronomy for twenty-seven years. Hanbury was the original 'boffin', so dubbed by a military man in the Second World War. Truly, that is where the name came from, and it was the title of Hanbury's memoirs. The line of research he pioneered was interferometry — how to measure the size and temperatures of stars by taking bearings from different starting points. I once took Hanbury to the History of Science Museum in Oxford. When I introduced him to the staff they went pale and tongue-tied, as you would if meeting Elvis, the Beatles and Barack Obama all at once. Another true legend. He once told Churchill to put out his bloody cigar as they entered a plane hangar full of fuel. Winston meekly obliged.

And a final hero from that department (there are lots): Ian Johnston once offered to do a *Science Show* series on the maths and physics of music. It was one of the best series we ever made, repeated often, and brilliantly produced by Mary Mackel.

Fiona and Fiona

There is a quaint adage, coming largely from disgruntled Easterners coveting gongs, that those wanting to be Australian of the Year should be called Fiona and come from Perth. Professor Fiona Wood arrived by bike for her *Science Show* interview. We sat on a bench near the University of Western Australia cricket pitch, at one of the loveliest university campuses in the world. She resembled a cheery mature age student with all the time in the world rather than what she is, a top-ranking skin surgeon with six children to look after.

Everyone in Australia knows that when a plane crashes in Indonesia, with dozens burned appallingly, their first stop when rescued, if they are lucky, is Fiona's hospital in Western Australia where she will operate around the clock. She may, in the process, use her famous artificial spray-on skin. Then, in her thirty-six-hour day, she teaches her students, some of whom I meet abroad (recently in Vancouver, for example) where they give her the highest praise and carry on an aspect of her associated research elsewhere. She is one of those whose reputation internationally is staggering.

Fiona Stanley is also an extraordinary mixture of relaxed charm and five-star accomplishments. She even has a large Perth hospital named after her, where Alan Bond died in June, 2015. The work that made her name was on folates, part of the diet essential for most pregnant mothers. She raised mountains of money (like Harry Messel) for both the research on children's ailments and their treatment through the 'telethon' method, which, indeed, is in the name of the Paediatrics Institute (the Telethon Kids Ward). She is committed to the welfare of Aboriginal children and has done much to make their lives more endurable — including providing a swimming pool in one of the settlements so that they can get good exercise, to marvellous effect.

I once was asked to do an ABC TV series, decades ago, on leading Australian scientists, which I called *The Uncertainty Principle*. I told the producer, Dick Gilling (former BBC producer on *Ascent of Man*, Jacob Bronowski's magnificent series) that Professor Stanley had to be included and could fly from Perth for the day. He was taken aback. *All the way from Perth? She'll be whacked.* And he'd never heard of her. 'Trust me,' I said, as I do so often, and Dick, dear heart, certainly did. Fiona, at the time, was a relatively junior doctor and convinced that she was her family's black sheep because the brothers were all eminent lawyers while she was a mere quack — on a level with 'trade'.

Fiona arrived at North Shore Hospital (given it was TV, you had to have a fitting backdrop!) and dived straight into our interview, gave us half an hour of television gold and then flew back. Dick was grinning like a minstrel. He put her on as the lead interview in the series.

Fiona is now on the ABC board and doing everything to help science. She is one of the truly great Australians.

Professors Lesley Rogers and Gisela Kaplan, University of New England

The University of New England in Armidale, New South Wales, is one of those, like Murdoch University in Perth and the University of Adelaide, that's had a shaky time at the top in recent years. But their science is still good. Plenty of ag and compost up there, and the stellar name of Professor Bridget Ogilvy, still remembered, who studied parasites at UNE before going to Cambridge. (Bridget later became head of the Wellcome Trust, did some innovative work with their investments and ended up being able to fund medical research to the staggering tune of £600 million a year, then equivalent to about $1800 million, equalling the UK government's contribution.)

My admiration goes to Lesley Rogers and Gisela Kaplan for their work on bird behaviour. Yes, I know birds have already featured heavily in this book, but they are so significant in Australia. Did you know, for example, that songbirds originated here? Tim Low recently wrote a book showing this. Lesley Rogers demonstrated that birds, like us, have lateralisation in their brains: left and right, doing different things. The effect is triggered by light impinging on the shell before the egg is hatched. One part of the brain concentrates on observation and will initiate warnings if, say, a hawk or other predator appears. Meanwhile the other part of the brain is concentrating on food and social matters. Lesley was elected to be a fellow of the Academy for her beautiful work. Gisela listens to birds and assures me that magpies can purr. I have never heard that but she says it's true, as is their imitation of some of our utterances.

> **Lesley Rogers:** Most of the work done up to fairly recently has been documenting the presence of laterality in animals, because there's a huge controversy. It used to be thought only humans were lateralised and now we know animals are, but are they more or less lateralised than humans? That fight's continuing. But what we're really turning our sights to now is to understand why lateralisation has been maintained across so many vertebrate species. And indeed there are even studies now showing invertebrates are lateralised. There was a recent paper showing that [the fruit fly] Drosophila, with an asymmetrical brain (structurally asymmetrical), can remember things.
> **RW:** What you're suggesting, to put it crudely, is a sort of division of labour in the brain so that you get more out of your brain than you might have by specialising.
> **LR:** Yes, exactly.
> **RW:** And that presumably would exist in most of the creatures in the animal kingdom with bigger brains?
> **LR:** Yes, certainly humans are strongly lateralised. The mechanisms by which it develops in humans might be quite different than in the chick. But there may be a similar kind

of selective pressure that's working on the brain to produce these types of individuals — some more lateralised than others according to predation.

Gisela Kaplan: I've actually recorded a sound of a magpie making an eagle alarm call in the presence of an eagle and it was quite different in structure. If I record that and then play it back to another group of magpies where there's absolutely no eagle about ... and I place the sound source on the ground then play it back at the same decibel level they would hear their normal calls ... the birds respond by looking up and scanning the sky instead of looking for the sound source. That's strong evidence that the bird is looking for something like an area predator rather for the sound source.

We've done that and it looks very promising. [Magpies] do that with the left eye, by the way. So there is a kind of fixed response every time to look up with the left eye, which means it involves the right hemisphere. It means it is absolutely the side that has already been established in other species to be the rapid response side for the detection of predators.

The Walter & Eliza Hall Institute

How can one centre have names of the quality of Don Metcalf, Jacques Miller, Gus Nossal, Suzanne Cory, Jerry Adams, and so on? I've asked this before and it's worth another look. The first two, Metcalf and Miller, should have won Nobel Prizes. But there comes a point with all mentioned above, that they have won so many other awards it ceases to matter. The mantelpiece is only so long.

Both Nossal and Cory have been presidents of the Academy of Science, and presented the Boyer Lectures on ABC Radio. The Institute has so many stars in one small galaxy.

One answer is the leadership of Mac Burnet. I've said, already, that he was a genius and demonstrated through experiment how the immune system works. New germs and old are constantly assailing your body, but being able to marshal a resistance to the new kind is the really fascinating miracle that the likes of Burnet worked out. *And he stayed in Australia.* Unlike Florey, Eccles and others, Mac committed himself to Australia long term. This allowed a team of critical mass to grow and give us the kind of centre of true excellence that is renowned everywhere.

Mac did have one hiccup with a relationship, though — not surprising when you consider the explosive nature of one Professor Carleton Gajdusek. Based in America but attached for some time to the Institute in Melbourne, Gajdusek wanted to go to Papua New Guinea to study the strange phenomenon of *kuru*, a 'laughing death' afflicting tribes in many villages. Against Burnet's orders, he went off there on a trip, studying local customs. He quickly decided (nothing slow about Gajdusek) that the fatal brain illness was caused by the preparation of corpses by the women, mainly, who picked up the virus or prion (a protein in the body) and passed it on, possibly through ceremonial cannibalism. Gajdusek's insight gave an explanation for any number of nervous ailments from CJD (Mad Cow disease) to scrapie in sheep. The work won him a Nobel Prize in 1976. I interviewed him many times and no recording was shorter than forty minutes. The relationship he built with villagers in PNG enabled the wild professor to import innumerable young boys, then girls, to his home in America and, eventually, he was charged with sexual interference involving some of them. Carleton Gajdusek is the only Nobel Prize winner, as far as I know, who has been to jail. He died in 2008. So maybe Mac Burnet was right all along. But the science, as befits the Hall Institute, was superb.

Gajdusek remains one of four guests on *The Science Show* to have spent time in jail. The others included two convicted of sexual harassment and one who served time for murdering a police chief in Canberra.

Peter Doherty

Peter is the only Nobel Prize winner, he tells me, to be trained as a vet. He is from Brisbane and did his ground-breaking research at the John Curtin School at the Australian National University in Canberra, with Rolf Zinkernagel — one of those situations where you put two bright guys in a small room and they either brawl or change the world. The pair chose to unravel an immunological mystery: how our defences actually identify the form of the invading germ so that a response can be designer-driven to make an effective reply. This is mediated through the T-cells. The result of the research was good news for transplantation therapy and meningitis, and much more.

Peter splits his time between the USA and the University of Melbourne, listens to *The Science Show*, appears on it, and writes lively books on everything from hot air balloons to *How to Win a Nobel Prize*.

Alan Finkel

There are not many independent individuals who qualify as scientific heroes outside CSIRO and campuses, but Alan does. He is certainly linked to noble institutions — he was the chancellor of Monash University, no less, and the president of the Academy of Technological Sciences and Engineering — but it is as a businessman and entrepreneur I celebrate him here.

Alan was an engineer specialising in nervous systems and also spent time at the John Curtin School. He developed technology that enabled scientists to 'listen' to individual nerves. This proved invaluable in labs all over the world. Sales of his technology were abundant and Alan was able to sell his company for a considerable sum, allowing him (when quite young) to think: *With all this money, why not have some fun?*

Instead of buying flash cars, a purple palace and the usual silly boys' toys, he set up a publishing firm called Luna Media and launched *Cosmos*, our premier science magazine (declaration: I am on the editorial board, with Buzz Aldrin and

Paul Davies). It was initially edited and run by Wilson da Silva and Kylie Ahern but now the able editor is Alan's wife Ella, herself the author of terrific books on genetics. The mag is now bigger and immensely attractive (with authors such as Alan himself and Norman Swan) and is making educational ventures into Australian schools. Alan and Ella are showing that you don't have to wait until the government has turned up to make a difference. At the end of 2015 Alan took over from Ian Chubb as Chief Scientist of Australia.

Mike Archer and John Long

These are both fossil men, but on a scale to make you feel faint. I have already mentioned Long's satellite chat with the dukes and David Attenborough involving his likely discovery of the first manifestations of sex (internal fertilisation) in vertebrates. But his career really took off at Gogo Station in Western Australia with an immense discovery of Devonian fish, about 350 million years old, with only the one at Canowindra in New South Wales, unearthed by a farmer and Dr Alex Ritchie, to rival it. John was always giving me a casual phone call saying, by the way, they'd just found a vast cave under the Nullarbor and had descended by rope (in the company of one poor cameraman) to find giant kangaroos and marsupial lions — was I interested? And, of course, it was soon written up in *Nature*!

John was formerly at the Western Australian Museum but then was tempted east, to Melbourne, then to run research at the museum in Los Angeles, but is now back in Australia and based at Flinders University in South Australia, where the discoveries keep coming. Amid all this shadowing of Indiana Jones, John also manages to write adventure books for children.

Mike Archer has been a museum man too, in Queensland, Western Australia and Sydney (at our oldest, The Australian Museum) where he was director. But his equivalent of Gogo is Riversleigh in northern Queensland, where the discoveries are so weird (his word) and large that helicopters and big trucks need often to be called in to fetch them. Mike's sense of humour goes as far as to name some of his ancient species after *Monty Python* — the authorities struggled to deal with this proposition. The work of Mike and his colleagues, with lots of citizen science involved, has opened up a history of giants and monsters in Australia. The diprotodonts are well known: the ancient marsupials the size of hippos. Then there are marsupial 'lions' and other carnivores showing parallel evolution with eutherian mammals (those with uteruses). The research also shows our Gondwanan history with many links to South American animals and even plants.

Mike may not write books for children (though he has done so for the general public) but his teaching is renowned. At the UNSW he regularly tops the favourite list, as determined by students' votes — not an easy achievement when combining rigour with wild metaphor. And, the best until last, he has an outspoken contempt for the kind of flaky 'science' I featured in an earlier chapter.

Department of Chemistry, University of Tasmania

One of my first ventures to Hobart included a visit to the university's chemistry department where Harry Bloom, a quietly spoken but formidably able professor, born in Palestine, told me of the frightening story of heavy-metal pollution in the river flowing through the city. It was both a horrifying and threatening revelation. It made the news when I put it on *The Science Show*. One young listener was Alexandra de Blas who promptly set her career on the path of remediation of the ghastly problem. She joined us in the ABC Science Unit after graduating to broadcast one of our environmental programs in 1976.

> **Harry Bloom:** I think there is a tendency by governments, and government instrumentalities, to hush up a lot of the facts that are actually available only in certain classified documents. The actual loss of mercury and other toxic

metals to the environment is well known in certain areas but the public just doesn't get to hear about it.

We have found that the so-called 'safe level' is only safe insofar as you eat a certain amount of fish per week. If you eat, say, 15 ounces per day — or actually, some people say 15 ounces per week — of the 0.5 parts per million contaminated fish, then there's nothing wrong. On the other hand, if you eat more than that you can develop some signs that may be associated with mercury poisoning.

Fish in Australia normally has only about 0.1 to 0.3 parts per million (ppm) of mercury but some large sharks have been analysed by various health authorities, for example in New South Wales, [and] have concentrations up to 15 ppm, which is about thirty times the so-called safe level. That is the reason for the banning of large sharks for human consumption.

RW: You've been quite vocal about heavy-metal pollution for a number of years and you've been called by various people 'irresponsible'. How do you answer that criticism?

HB: I've never been called irresponsible by scientists; I've only been called irresponsible by industrialists who, in many cases, have certain axes to grind and I'm afraid I can't take that sort of criticism very seriously. Most scientists who have communicated to me about my efforts have actually been full of praise rather than condemnation.

Nowadays the department does work across the chemical spectrum and I have been entertained by one of its present stars, Professor Paul Haddad, who does work on explosives and how effective our security at borders (airports, for example) may be. Without giving details, which he is not allowed to divulge, I remain aghast at the arbitrary nature of our techniques. No wonder the US system recently noted a failure rate near 90 per cent when they ran smuggled substances and weapons as tests through their very lax controls.

... And the museum in Launceston
It is called the Queen Victoria Museum and Art Gallery and the director is the son of John Mulvaney, whom I put in the top ten of Australian scientists in one of those polls in 2000, mainly because of his Mungo exploits and work putting Aboriginal history in a completely different light. Richard Mulvaney is doing great work with a delightful staff, showcasing science as something even very young people will relish. When I was there during National Science Week in 2014 and 2015, the place was buzzing and the youngsters were responsive and often struck silent in wonderment. It is thrilling to see what can be done in a regional centre if only you try.

They also have acres of old factory sites, pleading to be exploited as a growth centre. In two of the cottages on site they have set up a 'Hackers' Space' run by three or four young computer executives. They have crammed the rooms with equipment, old and new, and invited schoolkids to come in to learn programming. The youngsters not only learn advanced computer skills from professionals, they also design systems of use to the museum such as locomotive driving simulators. What's more the pupils are learning how to work in industry long before even leaving school. How brilliant is that? And the links with University of Tasmania are also excellent. That's where I reported earlier that undergraduates are winning international competitions for their engineering designs of drones, driverless boats and other self-propelling devices.

Australian Museum
Another declaration: I was president of the Oz Mus Trust in the 1980s and earlier 90s. Despite the cuts (thank you ex-Premier Greiner) we maintained the full complement of scientists and research. When the minister (my friend Peter Collins) complained about overseas trips he had to sign off on, I was able to say 'Minister, they are funded by outside money, not yours', and remind him that 60 per cent of research funds came from

sources other than the New South Wales government. The research coups of our second oldest Australian research centre (the oldest is the Royal Botanic Gardens, Sydney, about to mark it two-hundredth anniversary, having been founded in 1816) go back forever and involve the discovery of countless species, the recording of Aboriginal culture and anthropology, a succession of spectacular public lectures (from Attenborough, Richard Leakey, Jonathon Porritt, Sir Crispin Tickell — the ambassador who told Margaret Thatcher about climate change and had her committed support for its backing — Jane Goodall, and many more), those Devonian fish in Canowindra, and the collections, including one of the very best from all parts of the Pacific. The force driving these successes was Dr Des Griffin, one of the best museum directors (anywhere) of the twentieth century. The fact that he came from New Zealand and specialised in crabs helped in every way.

Then came more cuts and the depletion, as I recorded earlier, even of the vertebrate palaeontologists. Now a recovery is underway, led by the first female director Kim McKay, who pioneered Clean Up Australia with Ian Kiernan thirty years ago. There are young women there (Rebecca Johnson and Jodi Rowley) now revealing the koala genome and collecting new species of frog from Vietnam. I have often wondered whether Mike Archer's attempt to use DNA from the Tasmanian Devil to allow a cloned living thylacine to be created is still being worked on, but am too discreet to ask. The Oz Mus will always bounce back. It opened in 1827, the year Beethoven died, and is too magnificent to perish, whatever the forces of darkness try to do. Which is more than you can say about the Powerhouse Museum...

The Powerhouse Museum
I was there in the beginning with Dr Lindsay Sharp, the first director of the Powerhouse as we explored the old Ultimo Power Station (empty, apart from puddles, pigeons and the odd wrecked car) where the new museum was to be established. Then, in 1988,

it opened, with much vaudeville, acclaim and excellent funding, not least as a result of Lindsay's unstoppable efforts to gain private sponsorship. Then he resigned. Politics? Restlessness? He went on to run the Royal Ontario Museum (Toronto) and the great Science Museum in South Kensington, London.

After a while came the slide. Oh, the usual story: a state government having not much idea of how to foster a combined Applied Arts and Sciences Museum, and too fast a turnover of responsible ministers. Now there is a loony scheme to upend the lot and dump it in Parramatta. Can you imagine the expense of lugging all those planes and steam engines and original pre-Victorian pumping engines, going back centuries, to the West? Then there is the cost of converting the old building, designed for display, into ... what? An office, high-rise? Flats? Yes, you can make $400 million and more in real estate, but what is the point?

Lindsay Sharp and I have sent plans to Macquarie Street involving a new, exciting plan for Parramatta, involving all the Sydney institutions (Art Gallery, Botanic Gardens, both museums, State Library and Zoo), showing how they could combine to have shows for the twenty-first century, just as the Exploratorium in San Francisco did, and still does. About 15 per cent of both the Powerhouse's and the Australian Museum's collections are on display (the Powerhouse has a vast store out in Castle Hill in the north-west of Sydney) and they could be the basis for new exhibitions in Parramatta.

Have we had any reply to our proposal? Not a word. Former trustees, such as Trevor Kennedy, are said to be aghast. And the Vice Chancellor of the University of Technology, Sydney, Professor Atilla Brungs has deplored the plan publicly. As Professor Roy Green, head of business studies at UTS, reminded us on the *Science Show*: 'The district has a concentration of innovators to rival (almost) Silicon Valley. Why remove the CBD's only science museum now?'.

Climate Change Research Centre, University of New South Wales

Led by Matt England and Andy Pitman, with younger sparks such as Dr Donna Green, this is one of the centres constantly in the limelight. As I write, I see two letters from Matt in *The Australian* correcting a mischievous misstatement about southern ice. These naturally polite and reserved scientists constantly have to deal with belligerent attacks and personal comments as if they are agents for an invading power and not serious researchers of great repute. Have a debate, yes, but realise, as I've said before, that these are issues the public needs to think about, not a shouting match on a football field. Over many years these UNSW scientists have maintained the dignity of their science at the same time as producing publications based on excellent research. Their colleague at the University of Melbourne, Dr David Karoly does a similar responsible service.

Queensland Museum

Briefly, apart from all its splendid research on spiders and dinosaurs (such vast treasures), the Queensland Museum has now secured the New York-based World Science Festival from March 2016. This is a real coup for Professor Suzanne Miller, the director (a geologist), and will be less like a talkfest and more like the opening of the Olympics, with orchestra, drama via Alan Alda, and star appearances. Congratulations to Queensland.

Saint Peter's School, Adelaide

What is in the water there? Three Nobel Prizes from one high school? Robin Warren, who worked with Barry Marshall on *Helicobacter pylori*, the bacteria now linked to peptic ulcers. Before him was Howard Florey who led the team isolating penicillin as a drug in one of the most rapid and brilliant R&D operations in history, during war time, using bed pans as culture dishes. Before Florey there was Lawrence Bragg, and, it's worth repeating, the youngest person in history, still, to gain a Nobel

Prize in science. He was twenty-two when he did the work and twenty-five when he shared the prize with his father.

I always feel that Sir Mark Oliphant should have been at St Peter's, but he went to Unley instead. It did not inhibit him, however, and Oliphant became Lord Rutherford's right-hand man (Ernest Rutherford was rather clumsy) and achieved the highest results in physics, without scoring a Nobel. Sir Mark was one of the most principled and able of our scientists, concerned about disarmament. He was always regretful of the role he had played in the development of the atomic bomb and said so. Score one for Adelaide, if not Saint Peter's on this occasion.

Global Change Institute, University of Queensland

The University of Queensland boasts the Global Change Institute (GCI), directed by marine biologist Ove Hoegh-Guldberg (and I am on its board). Imagine studying oceans, fish, food, energy and coasts, and having a commitment to change them for the better actually written into your constitution. There has been limited change so far — mainly in the form of the GCI building, which displays all the energy-saving and modern green designs to help you have a zero carbon footprint, save water, eliminate dangerous chemical ingredients and much else.

It is from GCI's work that I learn that seagrasses are much more powerful absorbers of carbon dioxide, that there is a skill to putting solar panels on your roof (you don't just bung them up there) and that many power stations have been converted from gas and coal to solar. There is huge resistance to such work from instrumentalities. This came from Professor Paul Meredith in a long *Science Show* interview that was as promising in its outline of possible gains in energy technology as it was disappointing in its hope of real prospects of getting past the politics. What price change?

Paul, professor of physics at University of Queensland, specialises in organic solar as well. This was the junior player in the solar stakes — until now. Last year I was getting estimates of 9 per cent for the efficiency of solar materials you could print, like a newspaper. Imagine slapping this material on walls and curtains, even clothing one day, and having it pick up energy from light and converting into a state to be stored or transmitted. By contrast, solar silicon, once 27 per cent efficient in converting solar to electricity, now promises 40 per cent, according to Professor Martin Green at University of New South Wales. So now we have a real competition between the two main solar technologies, one stiff, familiar and commonly found on rooftops, the other more flexible, less powerful but capable of being spread like paint on almost any surface. Enter the perovskites.

Perovskites are materials containing lead, maybe one day tin (a safer metal) which have leapt to 20 per cent efficiency, thus *doubling* their power. If scientists, such as Paul and president of the Academy of Science Professor Andrew Holmes, can contain the lead and keep the organic solar material from damaging water, there will be a revolution on our hands. Or, more likely, on our buildings, roads, clothing, bags — anywhere a surface can be exposed to sunlight. Printing is cheap; slicing silicon is expensive. Who will win this race? That is real change.

Professor Steve Simpson
I met him first in Oxford where he was based at the Natural History Museum. Our interview featured diet and his discovery that many species, including our species, are satiated by protein. We eat any amount of accompanying fat and carbohydrate until that protein level is reached. This may be a main feature of the 'success' of the Atkins diet.

Steve is now at the University of Sydney continuing this splendid work, has been elected to both our Academy and to the fellowship of the Royal Society of London, and makes compelling ABC TV series on our landscape and agriculture, which have him parachuting out of planes and speaking to camera with all the flair of Brian Cox and Dr Karl combined. A real star!

Professor Tanya Monro

You may infer I have elevated the women too far in this book. Not so. Their record is extraordinary. Tanya Monro works on photonics or how you use light to communicate. Her detection systems can: 1. Test the quality of wine without having to open the bottle; 2. Monitor a baby's welfare at a distance; and 3. Assess the quality of materials in jet plane's fuel tanks. All this can be done by using light to detect the density and makeup of substances. The potential is boggling.

As a result of this R&D, Tanya has: 1. become elected to *both* Australian Scientific academies; 2. become Vice Chancellor Research at the University of South Australia; and 3. is engaged in a number of interactions with industry about promising applications.

And, by the way, she has young children, plays the cello and helps lead an orchestra. Any more quibbles about female power in Oz?

Lisa Harvey-Smith

Lisa is based at CSIRO in Sydney but travels all over to tell Australians about the SKA, the Square Kilometre Array now being built in Western Australia and South Africa. Her own field includes the evolution of cosmic magnetism and supernova remnants (you may recall that you and I are made of such remnants).

She comes from Britain and was attached to Jodrell Bank Observatory — the older sister of our Parkes Telescope. Other than her giving mesmerising talks on our (possible) astronomical future, she is, like Tanya Monro and her cello, an amazing achiever. She runs a marathon a day (often) and crossed the Simpson Desert, running, in six days. What does she think about on those enormous treks, I asked her. 'Food,' she replied.

Lisa-Ann Gershwin

Yes, that name does resonate. I could not help asking and was told that she is indeed of George Gershwin's family, a grandniece. Lisa-Ann is one of the world's experts on jellyfish and is attached to the CSIRO in Hobart. Her work when at James Cook University in Townsville, Northern Queensland, was on finding smaller relatives of the box jellyfish, which stops you swimming in northern waters in summer. These, the Irukandji, are barely 5 centimetres across and cause dreadful stings. She keeps discovering new species, as if to make our time in sea water in the north even more shocking than it was before. Fortunately she is now able to predict what conditions these mini box terrors will be found in. Watch out for north-east winds.

Stung! is the title of her bestselling book published by the University of Chicago Press. It has taken her on tours all over the world. In it she describes what will happen if we really do ignore warnings about depleting fish, doing nothing about acidification or warming, if we let things slide. Her last chapter describes oceans full of nothing but slimy pulsating jellies, some deadly, as their predators have been destroyed.

What should we do if that happens, I asked her. 'Adapt,' she replied.

And the rest?
There are plenty of stars in science in Australia. That's how my colleagues and I fill forty years with hot stuff and have lashings in reserve. There's so much to catch up with: Tom Maschmeyer's new nano lab at the University of Sydney; the perovskite revolution; the SKA being built in Western Australia; the innovation centre at University of Technology, Sydney; and any number of developments in private industry and those surprises you never imagine. Like CSIRO Physics in Epping who were looking for little black holes and developed the technology for Wi-Fi instead.

Can good government in Canberra or in our states afford to ignore, or give faint support to such a long list of talent? They have so far. Please let it be different soon. Our new PM has promised it will be.

But remember Lee Kuan Yew? — 'white trash of Asia'!

19 Being Co-opted Into the Elite...
When did excellence become an insult?

In the first chapter I brought in Lord Ritchie Calder with the amazing warning about climate change and how they had been worrying about it since 1963. I had forgotten this early gem until James Valentine played it for me on air (ABC Radio 702 Sydney) ten years ago as we celebrated another *Science Show* birthday. There was nothing in the statement that I had discounted in 1975 or made its memory fade thereafter. It was then just another important story and we did lots. It took nearly fifteen years for some of us to realise that the science of climate had become enormous and here was something we had to deal with as a priority.

I also mentioned that 'deniers' got put on routinely, until their cynicism and ideology became shameless compared with the research people who put out their facts and then tap-danced furiously to avoid going further with an opinion or a prediction. It was a tussle between purple propaganda and thoughtful evidence cautiously put.

Which is where Ritchie Calder's son Nigel comes in. He was, as I said, a BBC science documentary producer of note and I was prompted by his presence, and this was surprise, in the documentary *The Great Global Warming Swindle*. So I called him up at his home in England and asked whether we could record an interview about his resistance to the 'Swindle' and preference for the 'natural' explanation of a cause of warming perpetrated by cosmic rays. I opened the preliminary conversation with friendly gossip asking 'You live there in Sussex, quite close to my friends [both legendary science journos] Tim Radford, from *The Guardian* and Mike Kenward, ex-editor of *New Scientist*. Do you ever see them?'

'No,' he growled, 'I prefer to avoid the scientific establishment. Both have been co-opted and this makes their reporting unreliable.' I quote from memory, but he has written along these lines in his books. How had these famous communicators been compromised? I asked. By going to Royal Society receptions, joining organisations with scientific connections and, generally, being part of the Club, the priesthood, said Calder.

I then turned to the substance of the interview and registered my unwillingness to report on opinions I knew to be false. How can one ever justify spreading misinformation when you know it is wrong and there is not a hint of substance, within reason, to suggest an alternative explanation to the accepted scientific one? So, for example, I shall not say that women are stupider than men, that life exists on the moon, that Uri Geller is legit (except as a magician) or that there is no good evidence for global warming caused by people. There are journalistic ethics and, while the facts stand up as they do, I would be irresponsible to spread those ideas. Calder then erupted with the statement that he was as good a science journalist as me and gave a list of his attainments. At that point I began to wonder whether he had spent too much time in isolation.

There are two main considerations here: the first is whether I and my colleagues should eschew the champagne of the Academy's receptions, let alone their invitation to fellowships (I've been a Fellow since 1993), and secondly, whether the body of science is homogenous, like a church or the military, with one set of attitudes to which you become inured, whether you've noticed this or not.

Well, of course, both armies and churches have a wide variety of constituents with as many different attitudes as any other group of folk gathered under one umbrella. Was the military man T.E. Lawrence the same as Field Marshal Montgomery? Is Cardinal Pell the same as Father Frank Brennan? Is Fiona Stanley the same in her attitudes as Barry Marshall (no way!)? Science is far from a universal dump where robotic attitudes congregate.

I have been on the board of the Australian Museum, been an ambassador for the Powerhouse, the Queensland Museum, the Botanic Gardens in Sydney, on the board of the Science Media Centre, a fellow in Oxford (twice), a professor at UNSW, University of Queensland and a couple of other places (marginally). This gives me intimate contact with scientists high and low, with politicians of all colours, with prime ministers and with children. I remain friendly with them but will not hesitate to ask them the necessary questions on air that the listener would want answered; nor will I let them off the hook if they really should be taken to account. They would not expect otherwise.

In the early days, if I trod too carefully with a guest (such as Dr Helen Caldicott who campaigns against all things nuclear), I would be told, sweetly, how much more interesting it is to be given hardnosed questioning instead of an easy trot. It is true that one may be alerted to an issue by being around a board table, but all journalists depend on a range of prime contacts, tip-offs and leaks. If there is a conflict, I (we) say so.

Being away from all such influences has, pace Nigel Calder, the opposite effect to the one he averred. You can, through isolation, become fired up about a suggested miscarriage that is quite innocent, if only you had been in the loop and knew. So many stories in the media these days are made up of the herd, all running in the same direction with minimal glances either side to check the wider landscape (Depp's dogs, ice addicts gouging out their eyes, Kim K's bum) that it is essential to have a network that can help save you from avoidable embarrassment. So, for these reasons, I do not think we are co-opted, because science is not a monolith and being manipulated by clause 16 of the Old Mate's Act is rare. Being tied up in commercial or political considerations is another matter. Which brings me to the 'elites'.

The word 'elite' has become a pejorative term. John Howard uses it, as do the journalists and shock jocks who seem to write the same articles week by week (about the ABC having no right-wing presenters) or shout the same invective in radio shows to vast audiences. The idea is to convince voters, former stalwarts of Labor, that they have been abandoned by hacks who have no contact or understanding of the struggles of ordinary Australians. There are no more Chifleys or Curtins, goes the accusation, only smarty-boots graduates who whip straight into politics, have an aversion to the lumpen proletariat and spend their time hatching plots. This may be partly true, but the next bit of the 'elite' smear is to associate academics as a privileged class similarly separate from true Aussies, spending half their time with noses in the trough (the trough is funded by 'tax-payer money') or undermining the basis for Australian success and independence, that is the free-market and hard-working enterprise.

When, in addition to this attack on the campus and CSIRO, scientists turn up with worries about the environment and suggestions that enterprise be curtailed in areas such as mining, agriculture and some manufacturing, and you have the full case against this elitist conspiracy. The perpetrators are 'Watermelons' — green on the outside, red on the inside. This, at a time when no one of any consequence is pushing old-fashioned anti-capitalism and the likes of Jeremy Corbyn, Leader of the Opposition in the UK look simply silly. Attacking the straw men of these 'elites' has the advantage of first, making you believe everything is all right after all and, secondly, that you are not a dreary underachiever but a denizen of Struggle Street, abandoned by the silvertails.

The smear seems to have worked well, except that surveys appear to place scientists and medical researchers high in the trusted lists, and politicians and journalists way down low. Hence my question about whether the public really takes notice of these attacks on the so-called privileged 'elite'.

The maligned group of isolated lotus-eaters invariably includes the ABC, SBS and Fairfax. I have already discussed the ABC part of this supposed left-wing conspiracy but find the accusation of elitism as insidious. We are, in fact, through regional stations, phone-ins, comment sites and hard work to include outsiders, making intense efforts over decades to present the public's views on virtually everything. Discounting this by calling us 'elite', and therefore serving fewer Australians, can justify crippling cuts and possible dismemberment of the public broadcaster. Whatever the denials from the accusers, the results are clear: we really *struggle* (!) to bring our programs effectively to air. This book has repeatedly reminded you of the impediments. This is not a whinge but a statement of fact.

We are not, as a result, in a prime position to face the future, despite the efforts I have listed in other chapters. Paul Willis, once of *Catalyst*, has shown me what they are now doing at the RiAus (the Royal Institution of Australia, based in Adelaide in the same building as the Science Media Centre). Using a standard iPhone he can both film, edit and record commentary for professional standard 'film' segments to be used on their own channel or sent out to TV news departments anywhere. It is so cheap and flexible, especially when compared with the cost of a standard film crew. So why do we at ABC TV see unending evenings featuring British cop shows (spare me *Midsomer Murders* or I'll perpetrate one!) instead of the kind of rough but enthralling items that we once saw on *Hungry Beast* or the experimental young person's show run by Mike Rubbo, *Race Around the World*? This is the kind of TV that *can* be recorded easily in ways that Paul Willis's filming and editing technique exemplifies. What happened to that lively, fresh experimentation? I know it is on-line and on YouTube, but why not on ABC TV, with the authority that good journalism and careful production values can offer?

In summary, we have changes in society that I hope will not threaten the goodwill and decency I mentioned in another chapter following the Lindt siege and the Rozelle fire. When the population was asked, the good answer came. However, we are being buffeted by some unfortunate aspects of the new technology and the antagonism of special interests, though small, very loud. And there's the alteration of demographics such as the working class my parents came from, supremely interested in education and taking part in cultural events, now replaced by those, mainly men, who can no longer get jobs in manufacturing, nor even, perhaps, form a relationship with a future partner. Some say our working population contains about four million who are illiterate and/or innumerate (this was in the script of Dick Smith's ABC film on population). There are thousands left behind and open to hear it's not their fault but someone else's. Guess who? The intellectual elite? I think not!

This is summed up by Steve Hilton, former chief strategist for Conservative PM David Cameron. He writes in his book *More Human, Designing a World where People Come First*:

> Democracy is in crisis. It seems to serve the people no longer, but rather vested interests. Of all the bad that the vested interests do, perhaps their worst impact is the hold they have over our governments. It seems today that political legitimacy stems not from votes, but from money: the more of it you have, the more government pays attention to your concerns.

* * *

As for plans to continue *The Science Show* — I haven't got a clue. If we had some young staff it would be clearer. Is there enough science spread elsewhere on non-specialist programs? Well, apart from the thought that there never is enough science, what's different about our show is that we try to keep you informed of the main set of ideas currently coming from top-line research. A lot of what you hear generally is prompted by publisher's handy author tours; press releases from CSIRO and universities, often fair enough but more spoon-feeding; and blatant commercial spruiks disguised as something you really need to know.

Other programs such as *Future Tense* and *Background Briefing* can give you a thorough exploration of a theme: 'Will Artificial Intelligence Take Over?'; 'Can Robots Become the New Workforce?'; 'Will Your Phone Expose All Your Secrets To *Them*?' and other shouty subjects. I can rarely do that, but I can allow someone who wants to the space to have a go. I can also feature new talent by the score and respond to freelancers somewhere offering something fetching. What I can't do is commission a reporter to follow up a story — unless I just happen to have an intern from somewhere to answer the call.

So, *The Science Show* is a continuing experiment. It's a program of record, giving not just sharks and dinosaur news but some of the more arcane material (such as perovskites and new ideas about Ancient Greeks) you need to hear about. We can also set a standard for dense but listenable production — ideal for podding and calling up when you feel like it, whether it's a whole hour's worth or a single story, anywhere in the world. And we can obtain almost anyone you care to name: being forty means you know the crowd and even caught Brian Cox when he was a boy and not booked up like an opera singer with umpteen agents.

As we did forty years ago when Malcolm Fraser (my first science minister) as PM cut the ABC and kept me in Australia, we shall, as far as I know, blunder on (health permitting, of course). But even when I was being treated, in 2015, we did not miss a single *Science Show*. Our motto, borrowed from the dancing girls at the naughty review theatre The Windmill amid the bombs of World War II (who stayed in London regardless, as did the Royal Family), is *We Never Close*. Now that's tempting fate, isn't it?

> To the man-in-the-street who, I'm sorry to say,
> Is a keen observer of life,
> The word intellectual suggests right away
> A man who's untrue to his wife.
>
> W H Auden

Index

3D printing 188, 194, 196
7.30 175
60 Minutes 150
2007: A True Story, Waiting To Happen 55
2015 COP Summit (Paris) 95

ABC Rural 175
ABC TV 4, 5, 59, 139, 156, 175, 243, 256, 263–4
Abetz, Eric 85
Abrolhos Islands 93–4, 178, 212
Academy of Technological Sciences and Engineering 247
acidification 91, 136–7, 213, 214
Acquired Immune Deficiency Syndrome (AIDS) see AIDS
Adams, Douglas 109
Adams, Jerry 245
Adams, Phillip 18, 84, 162, 172
adaptation 222
ADHD 22
The Age 143
Ahern, Kylie 248
AIDS 22, 24, 33–6, 86, 102, 133, 213
  vaccine 35–6
AIDS and Australia 34
Aitken, Don 85
Albrechtsen, Janet 92
Aldrin, Buzz 247
aliens 18
All In The Mind 142, 156
Allen, David 155
Allen, Ian 156
Allen, Jim 198
allergies 225
AM 175
American Association for the Advancement of Science (AAAS) 35, 50, 108, 126, 176, 212, 217, 232
An Inconvenient Truth 84
ancient Greece 2
Anglo–Australian Telescope (AAT) 23, 33
The Animal Attraction 59, 60–1, 140
'animalcules' 225–6
animals 6, 55 see also by name
  antibiotics, and 223–5
  behaviour (ethology) 1, 55–67, 115, 140
  domestication 60, 61, 125
  evolution see evolution
  extinction of species 12–13, 14, 62, 121–3, 124, 131–2, 200, 205, 211–12, 228, 258
  human relationships with 58–9, 61–4
Anning, Mary 175
Antarctica 190, 194
antibiotics 221–5
anti-intellectualism 143
anti-psychiatry 5
apes 57–8, 120–3, 132–3
apnoea machines 69
Apollo missions 8, 28, 30, 31, 33
  cancellation of 7–8
Appleyard, Bryan 42
Archer, Mike 200, 214, 215, 248–9, 252
arms race 8
art 202–4
articulacy 174
As It Happened 73
asbestos 102, 104
Ascent of Man 243
Asia Pulp and Paper (APP) 149

267

Asimov, Isaac 109
astronomy 23–33, 107, 176, 190, 241
　citizen science and 176, 179–80
　radio *see* radio astronomy
astrophysics 127–8
ATMs 3
Attenborough, Sir David 56, 67, 136–7, 153, 156, 203, 209, 248, 252
Australasian Science 212
Australia 189–90
　goodwill in 234, 264
Australia and New Zealand Association for the Advancement of Science (ANZAAS) 107
　Congress 107
The Australian 49, 87, 92, 138, 143, 163, 254
Australian Academy of Science 21, 49, 52, 195, 239, 245, 256
Australian Conservation Foundation 21
Australian Museum 251–2, 261
Australian Sceptics 18
Australian Science Media Centre 153, 212, 225, 261, 263
Awakenings 120

Babbage, Charles 42, 154, 155
Bachelard, Michael 149
Background Briefing 20, 142, 146, 265
bacteria 219–22, 254
　children and 225
Bad Science 213
Baden Powell, Robert 147
Baker, Scott 214
Baldwin, Tony 98
Baliunas, Sallie 85
Baltimore, Professor David 35
Barbe Baker, Richard St 147–9, 154
Barenboim, Daniel 158
Barlow, Tom 38, 49
Barnaby, Dr Frank 8, 168
Barnaby, Wendy 168
Barnett, Professor Tony 56, 58, 102, 114–15
baseball scores 110
Battery Point 205

Beagle 3
Beaufort Bomber 71–2
Beazley, Dr Lyn 46, 186
Beddoes, Zanny Minton 49
Bell Burnell, Professor Dame Jocelyn 28, 29–30, 53
Bell, John 25
Bellybutton Biodiversity 177
Bernal, John Desmond 46
Betrayers of The Truth 12
Betty the Crow 1, 64–5, 144
Beyond Certification 149
Beyond the Mechanical Mind 18
Big Data 31, 188, 194
Big Idea 97
Bin Laden, Osama 82
biodiversity 32, 67, 190
biofilia 58, 59
biology 1, 24, 134, 164, 238
　marine *see* marine science
　molecular *see* molecular biology
　nerve 47–8
Bionic Ear 81, 186, 194
birds 6, 55, 56–7, 64–6, 136
　behaviour 243–5
　brains 65–6, 244
　citizen science and 180
　birth practices 9
Bishop, Arthur 71–3
Bishop, Bronwyn 107, 139
black holes 32, 70
Blackberry 80
Blackburn, Elizabeth 44, 187
Blakemore, Michael 83
Blatt, John 241
'Blood and Iron' 107
Bloom, Professor Harry 249–50
The Body Sphere 142, 156
Bolton, John 30
Bond, Alan 242
bonobos (pygmy chimpanzees) 57–8, 122, 123
Borgen 231
Borschmann, Gregg 146
boundaries, removing 185
Bowden, Tim 155
Bowler, Jim 199
Bowman, Alice 33
Boyer Lectures 245
Bragg, Lawrence 44, 46, 154, 254
brain-flux theory 96

brains *see also* neuroscience
  development 48
  fatal illness 246
  research on 113–14, 144
  salt and water 113
Brass, Dr Alistair 34–5
Breakfast 146
breaking the sound barrier 46
A Brief History of Time 25
Briggs, Dr Michael 22
British Foreign Office 9
broadcasting 155, 233–4
  art, as 156
  experiments 99, 102–4
  failures 138–46
  hoaxes 96–9
  innovation 99
  technology 166–7, 172, 263
  worldwide 218
Bronowski, Jacob 26, 243
Brooks, Geraldine 40
Brungs, Professor Atilla 253
Burka, Uwe 152
Burnet, Mac 96, 109, 155, 186, 245–6
bush-meat trade 122–3
Bush Telegraph 175
Butler, Harry 201
butterflies 149–50
Button, John 73–4, 77
Bygraves, Max 109
Byrne, Niall 173

Calder, Lord Ritchie 11–12, 259
Calder, Nigel 12, 85, 259–61
Caldicott, Dr Helen 261
Cameron, Clyde 162
Cameron, David 264
Campbell, Sir Philip 2
Canada 4, 63, 99, 162, 190
  mobile space arm 195
cancer
  astrophysics and 128–30
  cells 51, 129
  citizen science and 176, 177, 180
  dying of 102–4
  research 129–30, 176, 177, 180
car industry 72–3
carbon 137, 217–18, 255
carbon material and wiring 194
Carleton, Richard 150

Carleton, Sharon 43, 149–51, 154–5, 156
Carlsson, Ingvar 193
Carmody, Dr John 160
Carr, Bob 126, 162
Carson, Rachel 155
Carter, Bob 85
Castel Felice (Castle of Happiness) 157
Catalyst 32, 91, 102, 141, 156, 175, 215, 220, 263
cave paintings 203
CERN 31
Challenger space shuttle 83
Challis, Dr John 160–1, 163
Chandler, Jo 143
Charles, Prince 23, 98, 149–53, 155
cheese-making, mechanised 75
Cheltenham Science Festival 174
chemistry 187
chemotherapy 51–2
Chief Scientist of Australia 239, 248
chimpanzees 58, 64, 120–3, 132, 178
Chinese medicine 216
chloroplasts 222
Chubb, Ian 248
Churchill, Winston 241
cities 187
citizen science 94–5, 175–81, 216, 218
Citizen Science Association 176
Clarke, David 192
Clarke, Graham 194
Clarke, John 97
Clayton, Nicky 65
clean energy 90
Clean Up Australia 252
climate change 10, 11–12, 14, 32, 83–95, 141, 178–9, 229, 232, 234–5, 252, 255–6, 259
  citizen science and 178–9
  consensus, role of 88–9
  education and 87–8
  evidence 91–2
  local change, looking for 93–4
  opposition to 84–5, 134, 229–30, 232, 259–60
  scepticism, nature of 85–6
  wishful thinking 89–90
Climate Change Research Centre 254

clitoris 40–1
Clutton-Brock, Juliet 61
Clutton-Brock, Tim 59
coal 11, 81, 187, 214, 231, 255
coccolithophores 210
Cohen, Stanley 48
Coleridge, Samuel Taylor 3
Collapse 124
Collins, Peter 251
Columbia space shuttle 83
Colvin, Mark 20
Commission For The Future 84, 164
Commonwealth Scientific and Industrial Research Organisation (CSIRO) 53, 69, 70, 74–5, 85, 187, 237, 238–9, 247, 257, 258, 262, 265
communication
  interaction, and 144–6, 172
  satellites 195–6
  science in Australia 173
  technology and 170–1
computers 42, 172, 240–1, 251
condoms 130, 132, 133
confidence 53–4
Conniff, Richard 223–4
consciousness, nature of 113–14
consensus, role of 88–9
conservation movements 13, 21, 131–2, 166
'consilience' 91
conspiracy theories 140
contraceptive pill 22, 80
conversations 105, 172
Conversations 171
Coonabarabran 23
Cooper, Alan 200
Cooper, Caren 180
Copenhagen 83
Coppinger, Ray 60–1
coral 68, 136–7, 208, 209–10, 222
  bleaching 93
  mass spawning 211
Corbyn, Jeremy 262
Coren, Stanley 63
Cory, Professor Suzanne 53, 245
cosmology 32
Cosmos 143, 212, 247
Counterpoint 125
Cousteau, Jacques 153
Cox, Brian 67, 108, 256, 265

Crawford, Doug 167
Crick, Francis Harry Compton 46
Crossley, Dr Louise 116–17
culture, preservation of 202–4
Curie, Irene 44
Curie, Marie 44
Curiosity 6, 33
Cyclone Tracy 189

da Silva, Wilson 143, 248
Dagg, Fred 97
Daley, Michael 161
dark energy 31, 33
dark matter 31
Darwin, Charles 3, 110–12, 124
Darwinism 110–12, 124
Davies, Dr Paul 127–8, 248
Dawkins, Richard 55, 56, 133–6
Dayton, Leigh 143
de Blas, Alexandra 143, 145, 249
de Los Angeles, Victoria 158
de Waal, Frans 57–8
Deamer, Adrian 138
death, dealing with 22
Defence Science and Technology Organisation 196
deforestation 121, 146, 148–9, 151
Delingpole, James 92
Delroy, Tony 175
Demasi, Maryanne 102
den Hollander, Jane 186
denial 213, 259
  climate change *see* climate change
Denton, Andrew 182
Denton, Professor Derek 'Dick' 102, 113–14
Department of Chemistry, University of Tasmania 249–50
Descartes 56
Devonian fish 153–4, 191, 248, 252
diabetes 201–2
Diamond, Jared 109, 124–5
Diana indulgence 98–9
diet 17, 22, 242, 256
  Feingold diet 22
  indigenous 201–2, 204
  mental health and 220
digital technology 79
Dimbleby, Richard 98
Dixson, Alan 132

DNA 200, 222, 239, 252
Dodgeson, Professor Mark 69–70, 102, 195, 196
dogs 6, 140, 144
    domestication 60–4
Doherty, Peter 5, 247
Doogue, Geraldine 49, 101
Doolittle, Doctor 55, 67
Douglas, Roger 206
dragonflies 150–3
dreams 17
Drive 175
Drozdov, Nikolay 153
du Chatelet, Emilie 42
Dubai 229
Duckmanton, Sir Talbot 139, 160
Dunedin 206
Dutton, William 170
DVDs 3
dyslexia 22
Dyson, Freeman 85

Earle, Sylvia 209
Earthbeat 145
Earthwatch 178–9
Earthworm 145
Easter Island (Rapa Nui) 207
Eccles, Sir Jack 109, 154, 246
ecology 190
economics 10, 91, 112, 126, 206
The Economist 10, 49, 213
editing 100–1, 159
Ehrlich, Professor Paul 5, 108–9, 126–7
elephants 123, 130–1
elitism 262–3
Ellyard, David 143, 159
energy pathways 15–16
England, Matt 254
The Entrepreneurial State 69
entrepreneurs 69, 75
environmental challenges and issues 7, 21, 136–7, 164–6, 200, 229–31, 262
    deforestation 121–2
    innovation and 71, 76
    population 5, 13, 14, 126–7
epidermal growth factor 48
Epstein, William 10
ESP 15, 17
ethology 56–64

Eureka Prize 49, 141
Evans, Chris 182–3
Evans, Gareth 161
evidence 5, 32, 87, 91, 136
evolution 32, 59–60, 67, 109, 115, 124, 133–6, 144, 249
    attacks on 230
    blunders 115
    cosmic magnetism 257
    Darwinism 110–12, 124
    human society and 115–16, 124–5
Exploratorium (San Francisco) 253
explosives 250
exponential growth 164–5, 231
extinction of species 12–13, 14, 62, 121–3, 124, 131–2, 200, 205, 211–12, 228, 258

Facebook 171
faecal transplants 220–1
Faine, Jon 175
false-consensus effect 89
Famelab 174
Fanning, Ellen 168
Farrell, Professor Peter 101, 102
fear 235
feedback 176, 177–8, 181
    colleagues, from 105
Feingold diet 22
Ferry, Georgina 44
Fidler, Richard 171
Field, Professor Les 195, 196
Fields Medal 48
Finkel, Dr Alan 81, 247–8
Finkel, Ella 248
'fire-stick farming' 200
fires
    environmental damage from 217
First Australians 197
fish 153–4, 176, 212, 215–16, 248, 250, 252, 255
Fisher, David 100, 142, 154
fisheries 215–16
FitzRoy, Captain Robert 110
Flannery, Tim 84, 200
fleas 149
flexibility 235
Flores Island 203
Florey, Howard 109, 112–13, 155, 246, 254
Florey Institute 113, 186

fluoridation 17
focus and reflection 171
folates 242
forensic science 50
The Forest Trust (TFT) 148–9
forests 12
Forster, Alexis 50
Fossil Beer Can 97–8, 207
fossil fuels 11, 91
fossils 189, 191, 198, 248–9
Four Corners 175
fragmentation 171
Franklin River 21, 37, 198
Fraser, Malcolm 5, 265
Fraser, Professor 97
fraud in research 22
Frayn, Michael 83
Frazer, Ian 113
Free, Ross 181, 182
Fresh Science scheme 173
Freud, Sigmund 155
Friends of Science in Medicine 18
fringe material 5, 14
fruit fly 244
Fry, Peter 18
Fullilove, Michael 95
funding cuts
  ABC 3, 5, 142–3, 154, 160, 167, 171, 194, 263, 265
  Australian Museum 191, 251–2
  Powerhouse Museum 253
  public money, seeking 241
  renewable technologies 92
  science 181–3, 195, 206, 239
fungi 42–3
Future Fund 238
Future Tense 175, 265
futurology 13–14

Gajdusek, Professor Carleton 246
Galapagos Islands 2, 3, 12, 156
Galaxy Zoo 32, 176, 178
Gales, Dr Nick 215
Gallagher, Patrick 164
Gallo, Robert 35
Gallop, Geoff 193
Gandhi, Mahatma 117
Gandhu, Rajiv 153
Gardasil 113
Garrett, Kirsten 142
Gary, Stuart 32, 156

The Geek's Manifesto 183
Geller, Uri 18–19, 260
'Genesis to Jupiter' 107
genetically modified crops 230, 239
genetics 44, 115, 155, 164, 187, 238
  animal 59–60
  human migration and 207
  invented genes 115–16
  koala genome 252
genital mutilation 41
geology 187
geothermal energy 68
Germany
  science funding 183
Gershwin, George 257
Gershwin, Lisa-Ann 257–8
Giant Magellan Telescope 31
Gillies, Chris 178–9
Gilling, Dick 243
Gilmour, Dione 67
Global Change Institute (GCI) 255–6
global warming 12, 260 *see also* climate change
Global Warming Foundation 229
Gogo Station 248
Gold, Dr Julian 34
Goldacre, Ben 213
Gombe 121–2
Goodall, Jane 59, 64, 102, 120–1, 252
Google 170
Google Earth 195
Gore, Al 84
gorillas 122, 123
Goudie, Professor Andrew 189
Gould, Elizabeth 154, 155
Gould, John 155
Gould, Stephen Jay 109–12
Grade, Lew 158
Graham, Jory 22
graphene 186, 231
Graves, Jane 192
Grayling, A.C. 3
Great Barrier Reef 136–7, 179, 208, 209–10, 211, 212
The Great Global Warming Swindle 12, 84, 229, 259
Green, Dr Donna 254
Green, Professor Martin 70, 256
green alga *Spirogyra* 226
Green Chemistry 187

green policy 230–1
Greenfield, Baroness Susan 114, 153
Greenhouse '88 84
Greenpeace Australia 230
Greer, Germaine 39
Greiner, Nick 251
Griffin, Dr Des 252
guests 261
   changing nature of 168
   choosing 101–2
Guns, Germs and Steel 124

'Hackers' Space' 251
Hackett, Dr Earle 98
Haddad, Professor Paul 250
Hadfield, Chris 195
Halpern, Michael 232–3
Hamburger, Viktor 47
Handbury Brown, Robert 241
handshaking 225
Hanks, John 131
Harding, Professor Sandra 190
Harper, Andrew 59
Harvey-Smith, Lisa 257
Haskel, Jonathan 69
Hass, Hans 208, 209
Hass, Lotte 208
Hawaii 12
Hawke, Bob 68, 181
Hawking, Stephen 25–8
Hayden, Bill 150
Haydon, Tom 198
Hayne, Harlene 206
Hazelwood mine fire 217
healing 17
   laying on of hands 15–16
The Health Report 20, 21, 144, 220
Helicobacter pylori 254
Henderson, Mark 183–5
Hewish, Antony 29
Heyerdahl, Thor 14, 207
Higgs boson particle 31, 32
Highfield, John 4
Hill, Dorothy 109
Hilton, Steve 264
HIV 22, 35–6, 86, 102, 130, 213
hoaxes 96–9
Hobbes, Bernie 156
Hobbit, the 6, 203, 213
Hocknull, Scott 192
Hodgkin, Dorothy 44–6

Hoegh-Guldberg, Ove 209, 255
Holmes, Ben 192
Holmes, Dr Jim 103–4
Holmes, Jonathan 155
Holmes, Michael 192
Holmes, Professor Andrew 256
Homo micturans (Pissed Person) 97
hookworms 220–1, 225
hornbills 56–7
Horne, Donald 37
How Art Made the World 202
How to Win a Nobel Prize 247
Howard, John 139, 262
Hoyle, Professor Fred 23–5
Hubble Telescope 31
Hudson, Bob 97
Hudson Institute 13
Hughes, Howard 227
Hughes, Professor Terry 210, 212
Hughes, Robin 142, 161
human genome tracking 14
human reproduction 130–3
   sexual passion 40–2, 98
human society 115–16, 124–6
Humperdinck, Englebert 109
Humphries, Barry 96
Hunger for Salt 113
Hungry Beast 263
Hunt, Peter 98, 142, 143, 145–6

ideology 87–8, 93, 102, 259
ignorance, pooling 129
ILLIAC (Illinois Automatic Computer) 241
Illich, Ivan 17, 18, 34
'imaginary time' 26–7
immune systems 220–1, 246
immunology 186, 247
indigenous people
   art 203
   Australia 189, 190, 197–204, 207, 242, 251
   local, natural knowledge 201–2
   New Zealand 204–7
innovation 6, 10–11, 70–80, 175, 193, 236
   Australia, in 38–9, 68–70, 75–6, 95, 189, 238
   collaboration and 74, 77–80
   definition 79
   policy-making and 76–7, 258

Innovations 162
Innovations Institute (Wollongong) 188
Insight 96
intelligent design 134
interferometry 241
Intergovernmental Panel on Climate Change (IPCC) 84
International Space Station 195
International Union for Conservation of Nature (IUCN) 153
internet 3, 42, 93, 145, 170, 188, 212
interviews 100
  early technology 167
  editing 100–1, 159
Investigations 138–9, 161
Iraq War scepticism 85
Irukandji 258
isolation 261
Israel
  science funding 183
IVF 98

James, Henry 114
James, William 114
James Webb Telescope 31
Jameson, Graeme 81
Jarvis, Tim 7
jellyfish 257–8
Jenner, Edward 155
Jobs, Steve 69
Jodrell Bank Observatory 257
Johnson, Rebecca 252
Johnston, Emma 49
Johnston, Ian 241
Jones, Ann 142
Jones, Barry 37, 68, 84, 107, 164, 181, 182, 193
Jones, Caroline 155
Jones, Dr Rhys 197–8, 199, 202
Jones, Tony 4, 5
journalists 3, 101, 136, 155, 173
Jukes, Thomas H. 223
Jupiter 107

Kacelnik, Professor Alex 64, 66
Kahn, Herman 13
Kaplan, Professor Gisela 243–5
Karoly, Dr David 179, 254
Kaunda, Kenneth 131

Kelly, Fran 4
Kennedy, Trevor 253
Kenny, Chris 92
Kenward, Mike 260
Kerr, Professor Alan 239
Kew Gardens 239
Kiernan, Ian 252
Kimberley region 189, 194, 199, 203
Kimberley Rock Art Foundation 203
King Solomon's Ring 56
Kiruna 193
koala genome 252
Korea 76
Kramer, Dame Leonie 98–9
Kruszelnicki, Dr Karl 108, 156, 256
Kuan Yew, Lee 181, 258
Kulagina, Nelya 15
kuru 'laughing death' 246

Laing, Ronnie D. 9, 18
Lake District National Park 43
Lake Mungo 199–200, 251
Lamarr, Hedy 155
Lange, David 206
Lapland 193, 195
laptops 3, 42, 80
Large Hadron Collider 31
The Last Tasmanian 198–9
Late Night Live (LNL) 162, 172
Lateline 162
lateralisation 244
'laughing death' *kuru* 246
Laurie, Victoria 157, 190
Lawson, Nigel 229
laying on of hands 15–16
Lazaridis, Mike 80
Leakey, Louis 120–1
Leakey, Richard 252
Lehmann, Professor Nicholas 3
Levi-Montalcini, Rita 46, 47–8
Lewandowski, Stefan 85–90, 92
life, origins of 24, 128
light, series on 107
'The Light Fantastic' 107
Lindt Café siege 234, 235, 264
Lindzen, Richard 85
Lineweaver, Dr Charley 128
Lintott, Chris 176, 179–80
lithium for depression 70
Livingstone, Catherine 80, 238
Lofting, Hugh 55

Lombard, John 168
Lomborg, Bjørn 229, 230
Long, Camilla 134
Long, Malcolm 18, 162
Long, Professor John 153, 154, 248–9
Lorenz, Konrad 56
Lovejoy, Sir Clarence 96–7
Lovelace, Ada 42, 155
Low, Tim 244
'Lucky Country' 37, 181, 214
Lucy 201
Luna Media 247
Lyneham, Paul 4

McBride, Dr William 21
McClintock, Barbara 155
McGauran, Peter 139, 174, 183
McGuinness, P.P. 138
Machu Picchu 2, 156
McKay, Kim 252
Mackel, Mary 242
McTaggart-Cowan, Dr Ian 12
Magellanic Clouds 23
Mahoui, Iman 50–2
Malcolm, Lynne 142, 143, 156
Man Meets Dog 56
The Man Who Mistook His Wife For a Hat 119
MANIAC 241
Manning, Peter 138
Maoris 204–5
Margulis, Lynn 210, 221–2
marine science 49, 190, 208–18
Marohasy, Jennifer 85
Mars 6, 31, 33
Marsh, Helene 217
Marshall, Barry 155, 254, 261
Maschmeyer, Tom 187, 258
Mason, Peter 102, 106–8, 116
Massive Open Online Courses (MOOCs) 170
Masters, Chris 99, 155
materials 186
maths 107, 188
  music, and 241–2
Matthew, Delphine 157
May, Brendan 231
May, Fred 96–7
Mayes, Bernard 167
Mazzucato, Professor Mariana 69
Mead, Margaret 109

Medawar, Peter 115–16
media 93, 230, 261–2
  distortion 89, 92, 134, 234
  'game' 138–9, 162
Media Watch 140–1
Medical Nemesis 17
medicalisation of society 17
medicine 17–18, 21
  technological advances 69
megafauna 200, 205, 211
Meitner, Lise 53
Melanesians 207
Men of the Trees 148
Merck trials 36
mercury 231, 249–50
Meredith, Professor Paul 255
Merson, Johnnie 142
mesothelioma 102–4, 109
Messel, Harry 239–41, 242
metal smelting 75
Metcalf, Don 245
meteorology 76
methanol 231
microbiome 219
microorganisms 219, 221
migration 14, 207
military research 8–9, 195, 250
Milky Way 23
Miller, Jacques 245
Miller, Professor Suzanne 254
mineral resources 189
  mining in Sweden 193
  prospecting 76
Mirzakhani, Maryam 49
Mitchell, Natasha 143
Mitchell, Peter Chalmers 57
mitochondria 221–2
mobile phones 3
modernity 125–6
Mole, Adrian 182
molecular biology 46
Molomby, Tom 98
monkeys 57–8
Monro, Professor Tanya 257
Montagnier, Luc 35
Monty Python 28, 158, 249
Moore, Patrick 108, 176
More Human, Designing a World where People Come First 264
Morewell fire 217
Mornings 175

Morris, Desmond 115
Morwood, Mike 203
Mostyn, Sam 213
Mottram, Linda 175
Mount Stromlo Observatory 32
Mulvaney, Professor John 199–200, 251
Mulvaney, Richard 251
Mungo 'Man' (Woman) 200, 251
Munro, Dr Stephen 203
Murphy, Paul 4, 105
music
   maths, and 241–2
   Mozart on composing 119–20
myxomatosis 74

Nagra reel-to-reel tape recorder 166–7
'Naked Apery' 115
Nancy Millis Medal 49
narcotics 113
NASA 107, 195 *see also* space industry
National Broadband Network 79
National Geographic 230
The National Times 20
'natural philosophers' 3
natural selection 111–12, 133
   variation and 111–12
Nature 49, 64, 153, 154, 169, 187, 203–4, 248
Nature Climate Change 92
navigation, series on 107
Nazis and the role of iron 107
Neanderthals 6
neoteny 62
'Nerds and Losers' survey 181–2
nerve growth factor 48
neuroscience 52, 118–20, 154, 186, 247 *see also* brains
   neurology of the imagination 118
New Horizons probe 33
New Scientist 18, 141, 168, 188, 212, 260
New Society 162
New Zealand 12, 189, 204–7
   Maori 204–5
Newby, Dr Jonica 58, 59, 61, 102, 140–1, 214–15
Newman, Maurice 84, 87, 91, 92, 229

Newman, Professor Peter 193
Newton, Isaac 42, 111
Niblaeus, Kerstin 193
Nightlife 175
Nixon, Richard 7
no-take areas (marine) 210
Nobel Prizes 5, 23, 29–30, 32, 33, 35, 43, 44, 46, 48, 56, 115, 154, 155, 186, 187, 245, 246, 247, 254
Northern Territory 'intervention' 201
Norway 190
   science funding 183
Nossal, Gus 96, 245
Nuclear Research Foundation 240
nuclear weapons 10–11, 14, 125, 228
Nurse, Sir Paul 230

Obama, Barack 183
Ockham's Razor 96, 101, 102, 220
O'Connell, Dr Helen 40, 41
O'Dea, Dr Kerin 201–2
Off Track 142
Ogilvy, Professor Bridget 243
oil from algae 68
Oldfield, Barrie 147, 149
Oliphant, Sir Mark 96, 109, 255
open-records laws 232–3
ophthalmologists 145
overhunting 12–13, 62, 211
overpopulation 5, 13, 14, 126–7

Packer, Frank 239–40
Packer, Kerry 138–9, 239
palaeontology 176, 191–2, 252
Palmer, Clive 192
Papua New Guinea 204, 207, 246
parasites 220–1, 243
Parer, David 67, 156
Parer-Cook, Liz 67, 156
Parkes 23, 30
   Radio Telescope 28, 257
Parkes, Sir Henry 192
Parrots of Australia 156
past lives therapy 18
Pauly, Professor Daniel 215, 216
peace and war 125–6
Peacock, Dr Jim 238–9
Peacock, Matt 138, 168
penicillin 44, 45–6, 113, 254

penises 132–3
Perkin, William 155
perovskites 256, 258, 265
Peru 3, 14
Pettifer, Julian 153
PhD students 173–5
Philip, Prince 21, 153
photonics 257
Piltdown Skull 97
Pinker, Professor Steve 125–6, 145
Pinker, Susan 145
Pitman, Andy 254
Pittaway, Isabella 223–4
plagiarism 141–2
Plibersek, Tanya 160
Plimer, Ian 85
Pluto 33
PM 175
Pockley, Dr Peter 143, 159–60, 161
podcasts 99, 100, 101, 169–72
politics 230–1, 235
pollution 90, 165, 217, 249–50
Polynesians 207
population issues 5, 13, 14, 126–7
Poritt, Jonathan 252
Potter, Beatrix 42–3, 155, 156
Powerhouse Museum 252–3, 261
Poynton, Scott 148–9
precincts 194–5
Prime Minister's Science Prizes 80, 81, 188, 239
protein 256
psychokinesis 15
Pugwash 44, 46
pulsars 28–9
punctuated equilibria 109
pygmy chimpanzees (bonobos) 57–8, 122, 123

quantum computers 49, 187
quantum mechanics 31
quasars 28, 30
Queen Victoria Museum and Art Gallery 251
Queensland Museum 254, 261
Quilty, Professor Pat 215
Quirks & Quarks 163, 166

Race Around the World 263
Radford, Tim 260
radio astronomy 23, 28, 30, 53, 74–5

Radio National (RN) 125, 140, 146, 161, 168, 169
Radiolab 99–100, 171, 220
rainforests 179
Randi, James 18–19
randomised control trials 184
Rapa Nui (Easter Island) 207
Rasmussen, Lars 195
The Rational Optimist 228
recycling
   electronics 49
   tyres 49
Red Sea 208
reforestation 147–8
refugees 10, 44
Reid, Liz 39
relativity 31
religion 134–6
   science and 3, 109
remote sensing 195
requests for information 232–4
RiAus (Royal Institution of Australia) 91, 153, 263
Richardson, Professor Louise 49
Ride, Sally 109
Ridley, Matt 85, 228–9, 231
Rigg, Julie 162
Ringwood, Ted 109
risk 213–14, 232, 234–6
Ritchie, Dr Alex 191, 248
Riversleigh 248–9
Rizzardo, Ezio 187
Roberts, Alice 67
Robertsbridge 231
Robinson, Professor Carol 53
Rockwell, Norman 40S
Rogers, Professor Lesley 243–5
Roots To Riches 239
Rose, Stephen 5
Rosetta mission 33
Rothschild, Miriam 149–52, 155
Rowley, Jodi 252
Royal Botanic Gardens 252, 253, 261
Royal Institution of Australia (RiAus) 91, 153, 263
Royal Society of London 46, 49, 53, 230, 256, 260
rubber, history of 107, 116
Rubbo, Mike 263
Russell, Bertrand 44

Rutherford, Lord Ernest 255
Ryle, Martin 29

Sacks, Oliver 109, 118–20
Sagan, Carl 108, 166, 184, 221
Sahajwalla, Veena 49
Saint Peter's School, Adelaide 254–5
salt 116–18
   brains, and human 113
Sarich, Ralph 68
Saturday Extra 49, 175
Saunders, Alan 142
Schmidt, Dr Brian 5, 32–3, 52–3
science 2–3, 137, 162
   artists 156
   boundaries, removing 185
   communicators, types of 155–6
   continental scale 190
   jobs 182
   military research 8–9
   policy 38, 71–6, 143, 181–5, 190, 195, 258
   religion and 3, 109
   teachers 156, 183
science journalism 4, 91, 101, 136, 141, 143–4, 155, 156, 163, 260
The Science Show
   early ratings 17–18
   first broadcast 9–10, 162
   fortieth year 145, 174
   future plans 264–6
   origins 157, 161–2
   success, measuring 147
   succession planning 142, 143
   twentieth anniversary 120
Science Unit, ABC 7, 21, 98, 142, 157, 159, 175, 198, 249
Scientific American 223
'scientism' 5, 9
scientists 5–6, 108, 137
   defining 3
   fame 108
   role of 2–3, 32–3
   women 39, 42–54, 116
Scott, Dame Maggie 113
Scott, Mark 101
Scott, Ruby Payne 53
Scott, Sir Peter 131
seagrasses 217–18, 255
seahorses 216–17
The Selfish Gene 133

sex 153–4, 248
   human *see* human reproduction
sexual passion 40–2, 98
Shankar, Ravi 158
sharks 178, 250
Sharp, Dr Lindsay 116–17, 252–3
Shaw, Jenny 93–4
Sheil, Professor Margaret 53
Shirk, Jennifer 176–7
Short, Roger 130–3
Siding Spring Mountain 23, 33
Simmons, Professor Michelle 49, 187
Simpson, Professor Steve 256
The Sky At Night 108, 176
Sleepers, Wake! 37
Small Blue Dot project 166
Smith, Adam 112
Smith, Amanda 142, 156
Smith, Deb 143
Smith, Dick 18
smoking, lung cancer and 86
snails 109
solar power 68–9, 70, 231, 255
   organic 255
   silicon 255–6
Solomon, David 187
song-lines 202, 203
Soon, Willie 85
Southern Ocean 190
The Southwest 190
space industry 7–8, 31, 32–3, 76, 83, 190
   Australia 195–6
   NASA 107, 195
   Sweden 194
specialisation 143–4
species
   extinction of 12–13, 14, 62, 121–3, 124, 131–2, 200, 205, 211–12, 228, 258
   finding new 189–90
Speed, Terry 188
Spencer, Adam 108, 156
Spirogyra green alga 226
Spivey, Professor Nigel 202
spray-on skin 242
Sputnik 8
Square Kilometre Array (SKA) 31–2, 188, 257, 258
Stanley, Professor Fiona 242–3, 261
Star Stuff 32, 156

The State of the Tropics 190
statins 102
statistics 188
Steel, Harrison 195–6
stem cells 188
Stepenuck, Kris 177
Stockholm International Peace Research Institute 8
Stone Age hunting 62
storytelling 149, 202–3
Stung! 258
'sum over histories' 26
Sumner Miller, Julius 109, 156
'sunrise industries' 74
supernatural 18–19
supernovae 33
sustainability 165–6
　waste disposal 150–2
Suu Kyi, Aung San 190
Suzuki, Dr David 12, 153, 163–6
Svensmark, Dr Henrik 12
Swan, Norman 20–1, 22, 99, 102, 127, 142, 143–4, 156, 206, 220
Sweden 190, 193–4
　science funding 183

Tacon, Paul 203
Take Care program 122
talkback 99
tape recorders 166–7
Tasmania 189, 190
　indigenous people 197–9
TED 170
telecommunications 76
television
　1960s in Britain 158–9
　1970s in Australia 4
　colour 80
telomeres 44, 144, 187
Terania Creek 146
Tereshkova, Valentina 109
terra nullis 200
terrorism 49, 83, 195, 235–6
thalidomide 21
Thatcher, Margaret 46, 252
This American Life 171
Three-Minute Thesis Competition 173
Throsby, Margaret 161
Throsby, Professor David 161
Ticknell, Sir Crispin 252

Tiller, Professor William 14–17
Tinbergen, Niko 56
Tizzard, Dr Cath 205
Tonic 20
Top 5 Under 40 competition 174
Top End 189, 217
traditional owners
　protocols and 201
train travel 192–3
tropics 190
Trounson, Alan 98
Truganini 199
truth 5, 137
Turing, Alan 155
Turnbull, Malcolm 80, 234–6
Twitter 171
typewriters 172–3

ufology 18
Uganda 131, 178
The Uncertainty Principle 146, 243
underwater photography 209
United Nations Conference on the Human Environment (Stockholm) 7
United States of America
　science funding 183
universities 237–8

vaccines 17, 18, 69, 70, 102, 113, 230
　AIDS 35–6
Valentine, James 259
van Leeuwenhoek, Antonie 225–7
Vásáry, Tamás 158
Vaughan Williams, Ralph 148
Venus of Willendorf 202
Veth, Peter 203
Vietnam 4, 5
The Village Effect 145
Vincent, Professor Amanda 216–17
violence 125–6, 201
visual perception 118–19
vitamin B12 44, 46
volcanic ash 184–5
Voltaire 42
Von Däniken, Erich 18
von Frisch, Karl 56
Voyager spacecraft 107

Wade, Nicholas 22
Walker, Iain 89, 90

Walkley Award 140
Walter and Eliza Hall Institute 186, 245–6
war and peace 125–6
war on science 228–36
Warren, Robin 155, 254
water 210–11, 218, 227
   pollution 249–50
water-quality monitoring 177
Watson, James 46
wave power 68
Wayne, Professor Bob 59
wealth, creation of 6, 69, 74, 76, 175, 183, 238, 239
weather 170, 189, 228
   citizen science and 179
   climate change *see* climate change
   forecasting 110
The Weather Makers 84
Wellcome Trust 183, 243
Western Australia 190
wetlands
   destruction of 13, 151–2
whaling 214–15
Whetwell, Reverend William 3
White, Professor Peter 14
Whitlam, Gough 5, 39, 159
Wi-Fi 70, 75, 258
Widnall, Professor Sheila 53
wildlife 12–13 *see also* animals
Willesee, Mike 161
Williams, Pamela 158
Williams, Robin 120
Williams, Robyn 261
   arrival in Australia 157–8
   early career 158–60

Willis, Dr Paul 91, 156, 263
Wilson, E.O. 58
Wilson, Professor Margaret 205–6
wind turbines 213
wolves 59–61
Wolves of the Sea 156
women 39–40
   science, in 39, 42–54, 116
   sexual passion 40–2
   why men hate 39–40
Woods, Professor Fiona 242
wool industry 5
   science 75
World Safari 152
World Science Festival 254
The World Tomorrow 160
World Wildlife Fund 131
Wran, Jill 146
Wran, Neville 156, 167
Wright, Pansy 155
Wyndham, Arthur 161
Wyndham, Susan 161

X-ray crystallography 44–5

Year of Soil 219
YouTube 168, 170, 264

Zambia 131
Zen and the Art of Motorcycle Maintenance 27
Zinkernagel, Rolf 247
Zooniverse 180
Zuckerman, Al 26

www.ingramcontent.com/pod-product-compliance
Lightning Source LLC
Chambersburg PA
CBHW022039290426
44109CB00014B/913